'A fabulously entertaining read that explain
between impulsive pleasure-seeking and th
fication. *The Dopamine Brain* guides your understanding of self as you
ponder this daily tug of war. I loved it!'

Associate Professor Magdalena Simonis AM,
Department of General Practice, University of Melbourne

'*The Dopamine Brain* masterfully bridges the gap between complex neuroscientific concepts and everyday life, offering readers a clear and engaging exploration of how dopamine influences our behaviour. Dr Hronis provides knowledge, practical tools and techniques to empower readers to take control of their habits. I highly recommend it to anyone looking to recalibrate their life choices and live more deliberately.'

Dr Rachel Menzies, University of Sydney

'For the past forty-six years, I have worked at the coalface with a broad spectrum of criminals. From car thieves, drug addicts and mass murderers, the golden thread in the rich tapestry of these individuals' lives is dopamine. The dopamine rush they experience as a consequence of using or selling drugs, winning or losing on the punt, defrauding or injuring others, or merely getting away with a crime, inevitably drives and reinforces their self-destructive behaviour.

'*The Dopamine Brain* is an outstanding work, written in an eloquent and easy-to-understand manner. Dr Hronis has provided a succinct, easy-to-read outline and explanation of how this extraordinary neurotransmitter impacts upon so many aspects of our cognition and behaviour.'

Tim Watson-Munro, Criminal Psychologist

'This book is a must-read for psychologists, educators, academics, parents and anybody who is interested in learning how brain chemicals affect our behaviours. Written from clinical experience and a strong breadth of knowledge, Dr Hronis provides deep insight from the real world and offers strategies to positively affect individual wellbeing in everyday decision-making.'

Professor Marla Royne Stafford,
Professor of Marketing, University of Nevada

'Ever wanted to know what a good doughnut does to your brain? *The Dopamine Brain* is a brilliant book, written with authority, wit, and with lots of advice for a better life. Dr Hronis has read all the papers we all should have read and gives us a comprehensive introduction to the basics, the problems and the clinical applications of that strange neurotransmitter called dopamine. Her personal approach and the many (clinical) examples from her therapy room make this book a treat in several dimensions. A book that should be read by all.'

Professor Frans Verstraten, Former Head
of Psychology, University of Sydney

'*The Dopamine Brain* is a fascinating and engaging read. Not only did it provide me with a wealth of information on how this neurotransmitter is produced in the brain and how it affects thought processes, it also gave me the means to put this knowledge to practical use in my own life to more successfully live out my own values.'

Ted Hartwell, Executive Director,
Nevada Council on Problem Gambling

'Dr Hronis mixes science, clinical stories and even history in an intriguing story-like format about how dopamine is related to many things we do in everyday life, including gambling, drug abuse, excessive social media use and binge-watching TV. This book will help us find the balance between pleasure and purpose, using scientific theories of how the brain works.'

Professor Ahmed Moustafa, Head of School of Psychology and
Head of Discipline (Criminology and Counselling), Bond University

'Dr Hronis brilliantly distills the complex science behind our dopamine-driven behaviours into an engaging, practical guide for balancing short-term rewards with long-term fulfilment – a must-read for anyone seeking to optimise their habits and productivity.'

Dr Amantha Imber, bestselling author
of The Health Habit *and* Time Wise

The Dopamine Brain

Dr Anastasia Hronis is a clinical psychologist in Sydney, working across both academia and clinical practice. An academic at the University of Technology Sydney, she lectures and conducts research principally in the areas of addictions, intellectual disabilities and rural mental health. She travels nationally and internationally, presenting at conferences, as well as providing training for mental health staff around Australia. She is particularly fond of the time she spends working in rural communities. Anastasia is the founder of the Australian Institute for Human Wellness, which runs a clinic in Sydney where she and a team of psychologists provide psychological services to those experiencing mental health concerns. In addition to her work as a clinical psychologist, she is also a classically trained pianist and has performed as a soloist at the Sydney Opera House numerous times and at Carnegie Hall in New York.

The Dopamine Brain

Your Science-backed Guide to
Balancing Pleasure and Purpose

Dr Anastasia Hronis

PENGUIN BOOKS

UK | USA | Canada | Ireland | Australia
India | New Zealand | South Africa | China

Penguin Books is part of the Penguin Random House group of companies whose addresses can be found at global.penguinrandomhouse.com

Penguin
Random House
Australia

First published by Penguin Books, 2024

Copyright © Anastasia Hronis, 2024

The moral right of the author has been asserted.

Cover design by Alex Ross Creative © Penguin Random House Australia Pty Ltd
Cover image: Bogdan Dreava / Alamy Stock Photo
Index by Puddingburn Publishing
Typeset in ITC Berkeley Oldstyle by Midland Typesetters, Australia

Printed and bound in Australia by Griffin Press, an accredited ISO AS/NZ 14001 Environmental Management Systems printer.

A catalogue record for this book is available from the National Library of Australia

ISBN 978 1 76134 637 8

penguin.com.au

MIX
Paper | Supporting
responsible forestry
FSC® C018684

We at Penguin Random House Australia acknowledge that Aboriginal and Torres Strait Islander peoples are the Traditional Custodians and the first storytellers of the lands on which we live and work. We honour Aboriginal and Torres Strait Islander peoples' continuous connection to Country, waters, skies and communities. We celebrate Aboriginal and Torres Strait Islander stories, traditions and living cultures; and we pay our respects to Elders past and present.

This book wouldn't exist without the unwavering support of my mum and dad, who stand by and support me in all my endeavours. To my friends and colleagues who encouraged me along the way, thank you.

Contents

If the human brain were so simple that we could understand it, we would be so simple that we couldn't.
Emerson M. Pugh (physicist)

Author's note

Before delving into the pages of this book, I want to extend my gratitude to all the individuals who graciously shared their stories with such vulnerability and allowed me to document their experiences. They have taught me so much about life, about loss and about resilience. Having the opportunity to work with so many inspiring people and hear the details of their lives has forever changed my own perspective of what is truly important.

While the patients and stories discussed are based on real-life encounters, their names and certain aspects of their lives have been changed or deleted to protect their privacy. In some instances, a case example I provide is based on an amalgamation of a few patients whom I have seen for similar concerns.

Even though elements of their stories may have been changed, the emotions, challenges and insights they convey are deeply rooted in the lived experiences of real people. It is my hope that by sharing these stories in a respectful and sensitive manner, readers will gain valuable insights into the complexities of the human condition and find solace, inspiration and understanding.

May the stories contained herein serve as a source of reflection, empathy and connection.

The pleasure purpose dance

My clinic is on the fifth floor of a lovely office building overlooking the water near Sydney's Mascot airport. We're lucky to be close to the ocean so we get a great view, but we're far enough away from the airport that we don't hear too much noise from the planes. Aircraft take off and land continuously throughout the day and can provide quite the distraction from work. Gazing out the window, you can't help imagining flying off to some exotic holiday destination. We also sometimes use the selling point of 'come on in and watch the planes' as a way of enticing nervous children who've never seen a psychologist before!

In May 2023, something noteworthy happened in our neighbourhood. A new café opened, right at the foot of our office building. It was closer than the walk to our (previous) regular coffee shop, making it easy to pop down in between consultations. Over time, the baristas got to know our coffee orders by heart, making it an even quicker and more enjoyable process. This new café stayed open until 5 p.m., unlike the other coffee shops in the area that closed at 3 p.m. And not only that, the owner was an Italian-born pastry chef. Need I say more.

I've never considered myself to be a connoisseur of pastries, nor have I ever had a strong desire to eat doughnuts. Sure, I've enjoyed the odd Krispy Kreme. But I don't remember ever stopping at a petrol station or a supermarket specifically to buy one. I actually can't recall the first time I ate a Krispy Kreme, or how I felt during those first few bites. I can, however, tell you about the first time I ate a sugar-coated doughnut from the Italian pastry chef's café. The doughnut was freshly baked and still warm. The aroma was irresistible. Its golden-brown exterior was generously dusted with sugar. The outside was perfectly crunchy while the inside was soft and fluffy.

So just like that, literally overnight, I had a new favourite café. My team and I became regulars. So much so, that sometimes the café staff would reward our loyal business with free biscotti. We ate Sicilian French toast with mixed berry compote and ricotta, porcini truffle arancini balls and authentic, flaky sfogliatelle. (Note: Now is probably a good time to mention that I am in no way affiliated with this café, nor is there any incentive for me to promote their delicious food, other than the fact that they might read this and add some extra whipped cream to our Sicilian French toast.)

As the days went on, my desire for baked goods developed. I started craving something new, exciting and delicious. I'd feel it in the afternoon, usually around 3 p.m. I wasn't necessarily hungry. I wasn't short of food. In fact, I often brought in food from home, and I always had access to instant coffee at the clinic. But whether it had been a stressful day, a busy day, a slow day or a boring day, the circumstances were always just right for me to justify buying a snack. Suddenly, this Italian café served as a defining example of the power of a particular chemical in our brain: this was dopamine in action.

Dopamine – often dubbed the 'reward chemical' – is what our brain produces to reinforce a positive experience. I experienced a dopamine release when I ate that first sugar-coated doughnut. It told my brain and body, 'Hey, this is great! It tastes good and it feels good to eat it!' But dopamine does more than just tell us when something feels good. It also drives our behaviour to seek out that reward and pleasure in the future. It tells our brain, 'Hey, this is great, let's do it again!' My dopamine is thus activated not only when I'm eating the doughnut, but when I'm sitting in the clinic at 3 p.m. considering if I should to go down to the café for a mid-afternoon snack. It's even activated when I gaze longingly at the café's front display cabinet as I walk past on my way into work. What makes the pursuit of these particular doughnuts even more enticing is that they're always changing. Some days, they'll be filled with hazelnut cream. Other days, it'll be pistachio. Variety and unpredictability are exciting to the brain and activate the 'dopaminergic' pathways.

This is dopamine in action. It reinforces what feels good and drives us to seek out more of it. This can be wonderful, but it can also be dangerous. The 'dopamine drive' is typically quick, automatic and unconscious. It raises questions about how much choice and free will over our behaviours we really have. Are we truly as in control of our actions as we think we are? And what can we do to resist the pull of dopamine, which can be a very primal force?

•

Life is full of paradoxes. At times, it's complex, messy and chaotic. And yet it can also be simple, straightforward and ordered. It's both full of continual change, as well as stability and predictability. The weather varies from day to day, yet the seasons come and go without fail. Dynamics in relationships can change as time passes and yet love can be enduring. Our health is not guaranteed and we face unpredictable illnesses of all kinds throughout our life, yet we all grow old and death is inevitable. Life is never static and is constantly changing depending on our circumstances.

We can make deliberate choices to shape our life as we want it to be and yet we are at the mercy of so many factors outside of our control. We can be the captain of our own ship, steering it in the direction we'd like to go. We can choose our careers, decide where we want to live and select the people with whom we want to spend time. And yet, just like the captain of the ship who is at the mercy of stormy seas, we never *really* have full control.

But what about those behaviours over which we *think* we have control? How much of these are actually driven by *unconscious* processes? How much of what we do are we *actually* choosing, and how much of it is driven by processes within our mind and body of which we are not even aware?

It is a truism to say that the brain is incredibly complex. How it works has been a source of mystery and intrigue for thousands of years, having been studied by some of the oldest civilisations on the planet. In 1908, the German psychologist, Hermann Ebbinghaus, wrote that 'psychology has a long past, but only a short history'.[1] There is a lot of truth to this. Humans have long been interested in how the brain functions and the inner workings of the mind. However, the formal, scientific study of the brain – the field we refer to as 'neuroscience' – is a relatively young discipline. Ancient civilisations may have been interested in our grey matter, but they didn't have the knowledge or tools to be able to scientifically assess its function. It didn't stop them from trying, though!

The oldest medical texts we know of come from ancient Egypt and were written around 1800 BCE.[2] The ancient Egyptians are one of the oldest civilisations to have documented their extensive study of medicine. There are records of medical procedures such as setting broken bones, simple non-invasive surgeries, and dentistry. Keep in mind these were all written in hieroglyphics. Pretty sophisticated stuff for an ancient civilisation.

One of the oldest texts that specifically mentions the brain is the 'Edwin Smith Papyrus' circa 1600 BCE, so named after the American antiquities dealer who bought it in the nineteenth century. In the papyrus, the ancient Egyptians refer to the brain no fewer than eight times. For a science nerd like me, this is fascinating. They describe the symptoms, diagnosis and predicted prognosis of two patients with head wounds and compound fractures to the skull. A 'compound' fracture is where the bone is exposed and visible through the skin, unlike a 'closed' fracture where the bone is broken but the skin is still intact. A compound fracture to the skull would be considered a medical emergency today.

The descriptions also evidence some understanding of brain anatomy and function. There are references to 'pulsations of the exposed brain' and what we would now refer to as 'aphasia', which is the sudden loss of language (noted in the papyrus as 'he speaks not to thee'). Seizures ('he shudders exceedingly') are also mentioned.

There also seems to be some understanding of the 'laterality' of symptoms. What this means is that the Egyptians understood that the brain consists of two halves, or hemispheres, and that some bodily functions are specialised and controlled by one or other of the two hemispheres. For example, we now know that language processing and speech production largely occur in areas of the brain called 'Broca's area' and 'Wernicke's area', located in the left hemisphere. An injury to the left hemisphere can result in language being disrupted, while an injury to the right hemisphere is less likely to have the same result.

In ancient Greece, the work of Alcmaeon of Croton, a medical writer and scientist who lived around 500 BCE, further shaped our understanding of neuroscience. Alcmaeon suggested that the brain, not the heart, ruled the body and was the source of a person's intelligence. This had been the subject of longstanding debate among Greek philosophers and physicians.

Fast forward to the twentieth century and neuroscience began to be recognised as a distinct academic discipline in and of itself. We've come a long way from the days of the ancient Egyptians and Greeks. Nowadays, we have sophisticated neuroimaging techniques that can reveal intricate details about the brain and its activity. We understand neurogenetics and the ways in which specific genes are associated with neurological disorders. We understand neuroplasticity and the brain's ability to repair and rewire itself. Our understanding of the way in which neural networks function laid the foundations for the development of artificial intelligence (AI). Neuroscientists have even gone as far as developing technology that allows people who are paralysed to control robotic limbs and communicate with computers using their thoughts (also known as neuroprosthetics). All of which is to say, we've made extraordinary progress!

Biology, brain chemistry and genetics account for so much of who we are and what we do. Personality traits vary from person to person and are heavily influenced by our genes. Some of us are more extroverted, impulsive and open to new experiences, while others are more introverted, reserved and measured. These traits influence the choices we make in life – the type of career we choose, where we decide to live, and the people we select to have in our lives. But there's more than personality differences at play here.

Human beings are fundamentally wired for survival. This is why our reflexes cause us to pull away quickly when we touch a hot stove. Or why adrenaline can jolt us into 'fight or flight' mode when we're confronted with a dangerous situation. Brain chemicals like serotonin influence our moods. Melatonin influences when we sleep and wake. And dopamine drives us to seek out rewards like high-calorie foods and sex.

One of the challenges that modern-day life has accelerated is that of temptation. Temptations are all around and are easier to access than ever before. And when I say temptation, I'm not just referring to sex, drugs and rock and roll. I'm talking about anything desirable that seeps into the cracks of our daily lives. The temptation to watch one more episode on Netflix. The temptation to pick up your phone and check for new notifications on Instagram. The temptation to hit that snooze button in the morning, just one more time. And the temptation to eat another delicious, sugar-coated doughnut from the bakery downstairs from my clinic.

You've probably heard sayings along the lines of, 'short-term pain for long-term gain' or 'don't give up what you want most for what you want now'. We all face a constant tug of war between doing what feels good immediately and doing what will help us feel satisfied in the long run. When those two things align, *bingo!* What a success. But unfortunately, this doesn't always happen. Life requires us to balance what makes us feel good in the here and now with what will make us feel satisfied down the track. Getting that balance right is hard when the lure of instant gratification is ubiquitous.

If I want to buy something, I can jump on my phone, which is usually within easy reach, and automatically connect to high-speed internet, search online and find multiple products that suit my needs. I'm spoilt for choice. I pay for my item instantly, as my credit card details are stored in all my devices and all I need to do is hold up my phone and have facial recognition verify my identity. I can then have my item delivered to my doorstep within days, if not hours.

I no longer need to wait for the next episode of my favourite TV show to air on Netflix (let alone free to air). I can instead binge-watch an entire series with just one click of a button. Netflix and other streaming platforms will continue playing episode after episode, so I don't need to make an active choice to continue watching. In fact, if I really want to indulge, Netflix has a 'BingeWorthy' category that I can look at for inspiration.

If I'm hungry, I can contact one of the many online food delivery service platforms, choose from a vast range of options, and have a meal or treat delivered to my door within the hour. If I want to travel somewhere, a ride share vehicle will be at my location within minutes. If I'm bored, I can play a game on my phone, text a friend, check social media, read the news, or watch YouTube videos in the seconds that it takes me to swipe and unlock my phone or, even easier, have facial recognition unlock it for me! I can have Siri write me a to-do list, have ChatGPT proofread my manuscript (note to my editor, I did not do this!) and have Alexa adjust the temperature in my room – all with the tap of a finger.

But of course, you know all this because these temptations are likely a part of your everyday life too. We may not realise how effortless and automatic existence has become because it is so well integrated. And don't get me wrong, there are positives to all this automation. It saves time and money, improves safety, increases our accessibility to essential services and more.

But while the snooze button may help me feel good for an extra ten minutes as I'm lying in bed, it'll also make me feel stressed and rushed when I'm running late for work. And while the doughnut from the bakery tastes delicious while I'm eating it, if I ate one every time I was tempted, it would be disastrous for my health. Sleep ins are cosy and doughnuts taste great, but they don't align with my values or goals relating to my career or health. But should that mean I never have a sleep in and give up doughnuts for good? What fun would that be?

Herein lies the challenge of maintaining balance.

How do we balance pleasure and purpose in a world that seems designed to constantly tempt us in a hundred different directions? Amid the relentless busyness and distractions, how do we find a healthy equilibrium between fleeting pleasures and enduring meaning? And how does our brain respond to and navigate a world designed for constant consumption, where the greatest currency is our time and attention?

In this book, we'll unpack some key concepts relating to neuro-science in answering these questions. Specifically, we'll explore one tiny molecule in our brain that plays a huge role in driving our behaviours: dopamine.

The book is divided into three parts. In Part 1, we'll examine the evidence and research underpinning 'neurotransmitters', focusing on how and why dopamine and other brain chemicals work the way they do. I'll debunk some myths, explain what can happen when things fall out of balance, and outline how dopamine can lead to addictions. Why is this important? Well, the more informed we are, the better decisions we can make. By understanding how dopamine influences certain choices, we'll learn how to build greater behav-ioural control and make decisions which balance short-term pleasure with long-term purpose.

This is exactly what we explore in Part 2 – balancing the pleasure we can derive from dopamine-driven pursuits with meaningful and purposeful activities. We'll look at our values and how the 'dopamine drive' can get in the way of living a life aligned with those values. And we'll reflect on what's most important to us, to help guide us forward.

Part 3 is all about making those practical, behavioural changes. I'll walk you through the steps required to take a break from dopamine-driven activities, and provide strategies that you can use to build behavioural control over your urges. The goal is to help you live a life that bridges the gap between pleasure and purpose. Throughout, I'll also ask you to reflect on your own sources of ful-filment and how we can adjust the balance so that it's right for you.

So, let's get into it!

PART 1
UNDERSTANDING THE DOPAMINE DRIVE

In Part 1, we discuss the science behind dopamine – what it is, why it exists and how it works. We'll talk about what activates the dopamine pathways in the brain and how these subconscious processes drive us to make unconscious decisions. I'll debunk some myths like 'dopamine detoxes' and 'anti-dopamine parenting' which have attracted a lot of attention in recent years. I'll explain what happens when certain substances or behaviours activate the dopamine pathways too often and how the brain adjusts itself to accommodate. We'll walk through a range of activities such as shopping, gambling, drugs, alcohol, internet use, social media and even binge-watching TV shows, so you'll understand how dopamine works in relation to these behaviours. And finally, I'll ask you to choose one dopamine-driven behaviour that you'd like to change your relationship with. This behaviour will be our target as we run through a range of different strategies and activities in Parts 2 and 3 helping you to create long-lasting, effective change.

Chapter 1

What is dopamine and how does it work?

Given the volume of misinformation circulating online today, particularly on social media, you'd be forgiven for thinking dopamine is a trendy designer drug that will help you achieve all your goals. That couldn't be further from the truth! Dopamine has been around forever, but it was first 'discovered' in 1910 by the British chemists George Barger and James Ewens. It wasn't until the work of Swedish neuropharmacologist, Arvid Carlsson, in 1957, however, that we really came to understand the role it plays in the brain. Carlsson's experiments on rabbits showed that when he decreased the animals' dopamine levels, the neural circuits that influence movement stopped working properly. Later on, this was hugely influential in helping us understand the neurodegenerative disorder Parkinson's disease, where people develop problems with their movements as a result of lower levels of dopamine in the brain.

Since the work of Carlsson, research and interest in dopamine has expanded and developed to explore the role it plays in motivating us to seek out rewards and how it drives and maintains addictions. Research has explored the connection between dopamine and attention deficit hyperactivity disorder (ADHD),[1] as well as its presence (or lack thereof) in cases of depression,[2] and even how it relates to kidney function.[3] Based on the number of research publications about it, dopamine is arguably one of our most important and intriguing neurotransmitters.

What's a neurotransmitter? Well, it's essentially just a chemical 'messenger' in the brain. Neurotransmitters facilitate communication within the brain and the nervous system and are essential for maintaining both our mental health and physical wellbeing. There are many different neurotransmitters, each with a different function. Other than dopamine, you might have heard of serotonin, which is involved in regulating mood, sleep and appetite; Gamma-Aminobutyric Acid (GABA), which helps regulate anxiety and stress; and glutamate, which is involved in learning and memory processes. We discuss the role of various other neurotransmitters in more detail in Chapter 5.

Dopamine, like all neurotransmitters, contains the basic building blocks of life: oxygen, hydrogen, carbon and nitrogen. It exists not only in humans, but in all animals. Primitive lizards and reptiles living tens of millions of years ago had dopamine as part of their neurochemistry. What's more, dopamine serves a similar role in animals as it does in humans: it modifies behaviours by playing a crucial function in learning and reinforcement from rewards.

Dopamine is present in many different parts of the human brain, but it is especially rich in the prefrontal cortex.[4] This is the part where higher-level processing of information takes place and is largely responsible for planning, problem-solving, reasoning, creativity, information processing and working memory. It is also the reason why dopamine has a role to play in ADHD (more on this in Chapter 2).

One of dopamine's key roles is involved in the brain's reward system. More specifically, dopamine influences when and how we experience the feeling of pleasure and reward. But it does more than just help produce the experience. It also helps to motivate us to *seek out* pleasurable things. Therefore, it's released not only when we experience something pleasurable, but also in the *pursuit of* that pleasure.

Think about those sugar-coated doughnuts I mentioned earlier. Dopamine is involved in the enjoyable experience I feel when I bite into the pastry. But it will also then motivate me to seek out that tasty doughnut again (and again!). The next time I walk past the café, dopamine will fire up in my brain, reminding me of how good the doughnut tasted last time and creating a desire to enter the café and buy another one.

Studies confirm the role that dopamine plays in motivating us to seek out rewards. For example, dopamine neurons are stimulated *prior* to eating food, especially delicious food, but not so much during the actual consumption of the food, irrespective of whether it's delicious or not.[5] We see similar dopaminergic activity in rats, where dopamine is involved in the goal-directed actions of the male rat, such as approaching and mounting the female rat, and not just copulation per se.[6]

In addition to pleasure and motivation, dopamine does a few other things, including working with other neurotransmitters to control movement. It is present in the frontal motor areas of the brain – the basal ganglia – and is involved in motor learning and timing. Specifically, it helps us control voluntary movements (i.e. movements that we choose, like reaching out our arm to pick up a cup of coffee versus involuntary movements such as blinking). Low levels of dopamine have been linked to neurodegenerative disorders such as Parkinson's disease.[7] Depletion and disruption of dopamine results in the gradual loss of voluntary motor skills, including speech.

Dopamine is also involved in how we learn and how we remember things. It plays an important role in making sure that our memories are stored, and in consolidating the formation of new memories. Memories are essential to guiding our future behaviour. Recalling past experiences both consciously and unconsciously informs our future choices. If we bring this back to pleasure and reward and think about that tasty doughnut in the café, memory is vital. I need to *remember* that I ate that delicious pastry in order for me to have the urge to take the action to seek it out again.

From an evolutionary perspective, it makes sense that dopamine is involved in all of these areas: pleasure, motivation, movement and memory. They are all key to us seeking out a reward. For me to eat that pastry, I need to *remember* that it was delicious, have the *motivation* to seek it out again, *move* my body to return to the café and then experience the *reward* of it tasting great to repeat the process in the future.

Dopamine was crucial for the survival of the human species in the prehistoric eras. It motivated behaviours such as hunting for food and foraging. It reinforced the pursuit of finding shelter and safety and keeping away from predators. And it motivated people to seek out mates and to reproduce. In this way, dopamine has assisted with our basic survival. Without it, we wouldn't be here today!

It's important to understand the different ways that dopamine influences and shapes our behaviour, to see that it's not simply a 'feel good' neurotransmitter. One of its primary roles is in motivating us to seek out a reward, as well as providing the feeling of pleasure when we achieve it. In fact, dopamine motivates us to seek out and acquire rewards and goals even if they are not required for our immediate physiological survival.[8] This is worth keeping in mind as we consider healthy ways of accessing dopamine to feel good, but also using it so that we are motivated and feeling rewarded to do things that are meaningful and fulfilling.

Dopamine sometimes gets a bad rap because of the role it plays in addiction. It is so powerful that is has been colloquially referred to as the 'pleasure molecule' or the 'molecule of more'. It certainly can contribute to people forming unhealthy and addictive behaviours. We'll explore how pleasurable experiences can easily turn into addictions in Chapter 4. But it can also play a positive and useful role and there is a range of healthy ways that we can feel its pleasurable effects. For example, looking at photos and recalling happy memories helps us not only feel good, but also builds that desire to seek out those experiences again. Eating something that we enjoy feels great and it doesn't have to be a sugar-coated pastry! A healthy but tasty treat still gives us pleasure. Neuroimaging studies have found that dopamine plays a key role in us feeling good when we listen to music that we enjoy.

But here's the tricky part. I wish there was a formula that I could give you to help you strike the dopamine balance, like listening to music on your morning commute to work, eating something tasty for morning tea, or talking to a friend on the phone once a day. But it's not that easy. And that's because our experience of dopamine is relative to what has come before it. Let's break this down.

Our brain is constantly releasing dopamine at a baseline or 'tonic' rate. This is because dopamine is crucial to the functioning of our brain and body, irrespective of pleasure. Everyone is born with a different baseline level dopamine. In the same way that there are differences in personality characteristics, temperament, food intolerances and risks for genetic conditions, we also vary in our baseline levels of dopamine. The age-old debate of nature versus nurture is relevant here. Can we change our baseline level of dopamine or are we stuck with whatever we're born with? As with most nature versus nurture debates, the answer is 'yes *and* no'. While we are born with a natural level of dopamine, life experiences can certainly influence it.

Given that we all have a baseline of dopamine, our experience of pleasure at any given time is *relative* to our baseline rate and relative to what has come before. For example, if I play games on my phone all morning and get the dopamine release that that provides me, and *then* I eat something tasty for morning tea, I may not experience the same level of fulfilment or enjoyment that I would have, had I not played those games. The brain works hard to regulate itself and it won't allow us to be in a constant state of dopamine highs. For this reason, something which may have been a pleasurable experience, such as eating a doughnut for morning tea, no longer has the same effect because of what has come *before* it.

Sugar-coated pastries and other hedonic pursuits are okay. *In moderation.* The problems start when we indulge too often, too readily and without much effort. Our baseline tolerance for what is required to get that dopamine experience changes and suddenly, looking at happy photos isn't stimulating enough in comparison to the 'high' I can get from ordering something delicious on Uber Eats. Until very recently, finding a tasty treat required much more time and effort, hence the reward of eating it would have felt good and dopamine would have served to encourage someone to go through all that effort the next time around! Nowadays, the effort required is minimal.

Hopefully this chapter has provided you with a basic under-standing of what dopamine does and how it works. In the next few chapters, we'll unpack the 'pleasure versus purpose pursuit' even further, and investigate how our dopamine levels can become dis-rupted and what we can do about it. Dopamine can be powerful but, rest assured, it is not to be feared, but rather channelled.

LET'S REFLECT

- As we start to consider the role of dopamine in driving our behaviours and seeking out rewards, can you identify any instances in your life where it might have influenced your choices or actions? Consider my doughnut example. Is there anything in your life where you've felt the same pull?

- What are some of the natural, healthy things that you get dopamine from?

- Have you observed any patterns of dopamine tolerance? How do easily accessible pleasures affect your dopamine response over time?

Chapter 2

Debunking dopamine myths

Whether psychology is a 'science' or an 'art' has been the subject of debate seemingly forever. In reality, it's both. Psychology is always based on scientifically evaluated principles, but how we apply it to the work we do could be called an art form. This is because every person is different. Our biology, our life experiences, our genetics, the environment in which we grow up, our belief systems and the way all these factors interact differ from individual to individual. As such, the way we apply psychological principles to a person's unique situation must be nuanced.

Science is complex. The brain is complex. And neurotransmitters are complex. Explaining how a neurotransmitter like dopamine works isn't easy. Unfortunately, it can be simplified so much that the explanation loses meaning and becomes incorrect! Why does this happen? Well, for one, many scientific concepts require a deep understanding of specialised fields. Simplifying these concepts for a general audience can be difficult. What's more, scientific research often deals with nuanced findings that don't always neatly fit into simple narratives, and media (and particularly social media) tend to favour easily digestible stories. Now add to the mix headlines that sensationalise stories, social media influencers who explain (often poorly) scientific concepts, and our predilection for confirmation bias (that is, the desire to seek out information that confirms what we already believe), and we have ourselves a mixing pot for misinformation!

We witnessed this phenomenon during the COVID-19 pandemic, when the government, scientists and health experts had to explain how COVID-19 spread, why restrictions were necessary, how the vaccine worked, how clinical research trials were able to test its safety and efficacy quickly, and who was most at risk from both the virus and the vaccines' side effects. While there was plenty of good information around, there was also a lot of misinformation too (anyone remember the videos of people claiming that spoons could stick to the side of their arms after receiving the vaccine?).[1]

We've seen similar debates emerge out of complex science, with people claiming that vaccinations can lead to autism despite the evidence that this is not true.[2] Other people deny that climate change is real despite overwhelming evidence supporting it. Have you heard that goldfish have a three-second memory? I'm sorry to tell you, that's not true either. Goldfish definitely have a memory span lasting longer than three seconds, and marine biologists will tell you that goldfish are in fact pretty smart![3] How about this one: did you learn about the 'tongue map' at school? That different parts of the tongue are receptive to different tastes such as sweetness, bitterness and sour tastes? That's also false. The receptors that pick up these tastes are located *all over the tongue*. The scientific community has known this for a long time, but somehow the story that tastes are segmented was widely spread, perhaps because it was simple and easy to understand and communicate.

This chapter is all about correcting some of the myths you might have heard and might even believe about dopamine, all the way from what a 'dopamine detox' is to the role dopamine plays in addictions.

MYTH: A person can detox from dopamine

FACT: Wrong! You can *never* detox from dopamine. Detoxification, or 'detox' for short, is when you eliminate a chemical from your body. When someone abstains from a substance, the body is able to rid itself of that particular toxin or unhealthy substance. For example, if someone goes through an alcohol detox, they stop drinking and allow the body to rid itself of alcohol-related toxins. A detox is at best unpleasant. At its worst, it can be life-threatening. Withdrawal symptoms are common and can vary from mild to severe, depending on how much an individual drank, how often they drank, and if they have any co-occurring conditions.

In the context of dopamine, a detox is impossible. Dopamine is naturally occurring. It plays a significant role in various aspects of human physiology. As we've seen, it's involved not only in the pleasure and reward centre of the brain, but in motor control, motivation, arousal, memory, sleep and executive functioning. If we completely detoxed from dopamine, we wouldn't be able to function let alone stay alive! In fact, our body biologically won't allow us to detox from dopamine, as it is naturally and automatically produced.

When the idea of a 'dopamine detox' became popular, it was unfortunately taken to mean something different from what was intended. A dopamine detox was meant to refer to detoxing (or avoiding) the *behaviours* and *substances* that activate the dopamine pathways in the brain – for example, abstaining from social media, technology, or alcohol for a period of time. The 'dopamine detox' was misinterpreted, spread through social media and then promoted as a lifestyle and productivity trend in Silicon Valley in California, which further added to its appeal.[4] Later in this book, we'll talk about healthy ways to detox from the substances or behaviours that are excessively activating your dopamine reward pathways. We'll also look at how we can build behavioural control in evidence-based ways.

MYTH: You can become addicted to dopamine

FACT: Wrong! You *cannot* become addicted to dopamine because it's not inherently addictive. It is a chemical messenger that helps us pay attention and repeat certain actions or experiences. What really happens is people become addicted to *activities* or *substances* which produce dopamine. When we constantly engage with something that produces dopamine, like drinking or gambling, for example, the brain becomes conditioned to seek out more and more similar experiences in order to receive more and more dopamine. However, the brain also becomes used to the dopamine circuit being activated, meaning we need more and more *intense* experiences in order to feel that same level of excitement. In this way, we become addicted to what produces the dopamine, rather than the dopamine itself.

MYTH: The more dopamine you have the happier you are

FACT: Not quite. The relationship between dopamine and happiness is more complex than a simple formula of 'more dopamine = more happiness'. Dopamine is involved in the experience and pursuit of pleasure. However, if we continuously seek out activities and substances that lead to a surge in dopamine, it results in 'tolerance'. This means that the same level of stimulation – from drinking to gambling to excessive social media posting – no longer produces the same level of pleasure.

Our brain naturally wants to remain in a state of 'homeostasis', that is, in a state of internal balance. It therefore won't allow us to consistently remain on a dopamine 'high'. Tolerance will build and our brain will change the way it produces dopamine to regulate our experiences. This is where we run the risk of slipping into addiction, as the hunt for more stimulation can become compulsive.

MYTH: Dopamine is solely responsible for happiness

FACT: Again, false. Happiness is a complex emotional state influenced by so much more than just dopamine. There are multiple neurotransmitters that are involved in the process of regulating mood and emotions such as serotonin, endorphins and oxytocin. Beyond this, our circumstances, our life experiences and our biological emotional sensitivity all play a part in how we perceive happiness.

MYTH: Boosting dopamine levels can improve focus and productivity

FACT: This needs to be considered carefully. Remember how we said science is complex, people are complex and there are lots of nuances? Well, here's where that applies! Both too much and too little dopamine can negatively affect our focus and productivity. While an increase may enhance focus and productivity for some tasks, it can also have the opposite effect, resulting in distractibility, impulsivity and a decrease in cognitive regulation. Individual factors such as genetics and overall health influence how a person responds to changes in dopamine levels. Balance is key and a one-size-fits-all approach does not apply here.

Dopamine is said to play a key role in the condition ADHD. In fact, in telling people I was writing a book about dopamine, the two most common responses I got were, 'Have you heard about dopamine detoxes?' and 'I have ADHD. I can be your case study!'.

ADHD is a neurodevelopmental condition which is characterised by persistent patterns of inattention, hyperactivity and impulsivity. These features affect day-to-day life for people with the condition. ADHD is typically diagnosed in childhood, although in recent years there's been an increase in adults being diagnosed.[5] There have been a number of theories as to why, with some experts worrying about the risk of over-diagnosis, and the corresponding over-prescription of stimulant medications. However, it is also likely that as awareness and healthcare have progressed, so too has the diagnosis of adults who potentially struggled with learning, emotional regulation and socialisation throughout childhood.

Since 1999, studies have examined the role dopamine plays in ADHD. Back then, researchers found a 70 per cent increase in dopamine transporter density (or DTD) in adults with ADHD, compared to those who didn't have ADHD.[6] Higher DTD results in less dopamine, which might be a risk factor for ADHD. Since then, there have been many more studies that consistently show a link between DTD and ADHD. DTD certainly doesn't always indicate ADHD, but it may be a useful screening measure.

The main medications prescribed to treat ADHD are stimulants such as dexamphetamine, methylphenidate (sold as Ritalin or Concerta) and lisdexamfetamine (sold as Vyvanse). These medications work by increasing the amount of dopamine and noradrenaline in the brain by blocking the transporter that prevents their re-uptake and rapid removal. This occurs in the prefrontal cortex, the part responsible for our executive functions such as attention and impulse control. While stimulants do not help with all ADHD symptoms, they reduce the symptoms in about 70 to 80 per cent of adults.[7]

Stimulants can have a range of side effects, which is why their prescription is tightly regulated. There is potential for abuse, as they can also create a sense of euphoria and contribute to weight loss. It's a common myth that stimulants improve the concentration and productivity of people who don't have ADHD. A group of forty people were given online arithmetic tasks to complete during four sessions.[8] At each session, they were either given a placebo or a stimulant before completing the task. The results showed that while the stimulant did not affect getting the correct answer, it did increase the number of moves and the time spent solving the problems compared to the placebo. Taking the stimulant medication actually resulted in a *reduction* in productivity!

The myth has likely prevailed because people taking the stimulants *do* feel different. They are, after all, taking a medication that speeds up the messages travelling between the body and the brain. People may 'feel' more alert, but this does not necessarily mean they are more productive.

MYTH: A 24-hour 'detox' from activities or substances that create dopamine spikes can reset our dopamine levels

FACT: False! Don't believe everything you hear on social media! Again, we have a case of oversimplification, perpetuated by social media influencers. Twenty-four-hour 'dopamine fasting' is a lifestyle trend that has become popular in Silicon Valley, where people cut themselves off from almost all stimulation for a day.[9] Through this process, their aim is to abstain from actions that are associated with hits of dopamine and 'reset' their dopamine levels. It is true that excessive and chronic stimulation of our dopamine system can lead to desensitisation of dopamine receptors and a reduced sense of pleasure (what we refer to as building a tolerance). However, the answer to this is not as simple as a short 24-hour detox from certain activities.

In reality, dopamine regulation is a complex process influenced by many factors, and it doesn't undergo a sudden reset in a short 24-hour period. In fact, it is almost impossible to detox from everything that produces dopamine. Dopamine can come from small but pleasurable things like eating something that tastes nice, hearing music that you enjoy or spending time with a good friend. This isn't to say that abstaining from actions associated with dopamine for twenty-four hours can't be a useful thing. But let's be clear about what's actually happening. A 24-hour 'detox' ends up being more about experiencing the discomfort of withdrawing from things like checking social media or eating sugar, and building behavioural control over the urge to return to that behaviour or food. In fact, it's barely that. Twenty-four hours is only long enough to distract ourselves through some cravings or urges. It's not long enough to make long-lasting, meaningful change. Unfortunately, that takes time and is a tougher process. In the case of addictions, it takes far longer than twenty-four hours to reset both the level of homeostasis in the brain and our dopamine levels.

It's unrealistic to expect change to happen too quickly and dramatically. As we'll see in Chapter 7, it's better to make small changes, which are more easily implemented into our day-to-day life, rather than drastic changes that are likely to be unsustainable.

MYTH: 'Anti-dopamine parenting' is an effective way to raise children

FACT: Whenever I hear phrases like 'anti-dopamine parenting' in the media, I get really nervous about what we're being sold. This feels like a misinterpretation of science simply waiting to happen. Just like people started to believe that you could detox your brain from dopamine activation (which as we've said, is impossible to do), I worry that parents will start to believe that depriving their children of technology or stimulation will be effective in reducing their dopamine activation.

The essence of 'anti-dopamine' parenting is the reduction of excessive triggers of dopamine activation. Essentially, it's the activities that give children instant gratification; for example, watching their favourite TV shows, playing video games, playing on the iPad or using social media on their phone. The approach is not fundamentally bad. I'm all for parents wanting to limit and manage the amount of screen time their children's developing brains are exposed to. In fact, in 2024, the US state of Florida brought in tight controls to restrict children's social media use. From 2025, children under fourteen will be banned from using social media, and those aged fourteen to fifteen will need their parents' consent to join a social media platform.

Too much screen time has been associated with all sorts of problems, such as obesity, irregular sleep, insufficient consumption of fruits and vegetables, disordered eating and an increased sedentary lifestyle.[10] Between 45 and 80 per cent of children in fact fail to meet the international recommendations of less than two hours of screen time per day.[11, 12] So, the anti-dopamine parenting 'hack' comes with good intentions.

What concerns me is the phrase 'anti-dopamine'. It suggests dopamine is a villain we are trying to avoid or banish. Articles on anti-dopamine parenting will tell you 'it's not the child you're fighting, it's their dopamine', with dopamine portrayed as an evil and addictive chemical wreaking havoc in the brain. Again, this is where we run the risk of oversimplifying science. Sure, dopamine has a role to play in kids' desires to watch another episode of *Bluey* or play more *Minecraft*. But let's not forget that their brains are not developed in the way that adults' brains are. Children are more impulsive and their ability to regulate their behaviours and emotions is still forming.[13] 'Anti-dopamine parenting' is misleading. Perhaps it should instead be called 'moderated video game parenting' or 'reduced technology parenting', but that's not quite as catchy. Really, this is just a new label stuck on the notion of 'setting boundaries'. Dopamine will still play a role in a child enjoying a tasty meal, playing with friends at the park or going to a music lesson.

Why would we want to be 'anti-dopamine' and deprive our children of all joy and pleasure in life, anyway? Or worse, teach our children that life is not to be enjoyed? If we want to help them grow up to be effective adults, we must teach them that life is filled with both pleasure *and* pain. We don't need to deny ourselves all pleasure. We need to show our children how to regulate their desires and impulses and manage the behaviours, foods and technologies that might tempt them to consume more.

LET'S REFLECT

- What myths have you previously believed about how dopamine works?

- Have you come across any examples of misinformation spread via the media or social media?

- How has your understanding of dopamine changed?

Chapter 3

What can disrupt our dopamine?

Now that we've taken a look at what dopamine is and how it works, let's consider when things can get out of balance. Dopamine is incredibly useful and important for motivating us to seek out certain activities, behaviours or outcomes. But it can also lead us towards harm.

Over time, our body builds a tolerance to the stimulation we receive from the outside world. This means that instead of having constant highs and dopamine hits, the body seeks stability. 'Homeostasis' is the brain maintaining a stable and balanced internal state, ensuring proper and effective functioning. If we repeatedly expose ourselves to substances or behaviours that release large amounts of dopamine into the brain's reward pathways, we can change and lower our baseline level. Our brain then tries to compensate by naturally producing and releasing less dopamine.

From an evolutionary perspective, it is unnatural for us to be constantly stimulated by delicious treats, games on our phone, social media notifications or endless content on Netflix. In some ways, modern life is like Disneyland for our brain. But think of it this way. The first time you visit Disneyland, you're excited and overwhelmed.

There are vibrant colours, dazzling lights, classic Disney tunes coming from the carnival rides, and boundless possibilities that provoke a sense of wonder and excitement. Now imagine that you *work* at Disneyland. Exposure to the sights and sounds is a part of your routine. Eventually, the bright lights and melodic tunes lose their lustre, fading into the background as familiarity sets in. In fact, you'll probably feel irritated hearing 'Let It Go' or 'Under the Sea' for the thousandth time. It would be way too stimulating to experience that same high every time you walked through those big, inviting front gates. It'd be hard to do your job because of all the distractions. In the same way that you'd 'adjust' to working at Disneyland, your body and brain adjust to stimulation. And because of the brain's desire for homeostasis, our baseline dopamine level adjusts. We now need something *more* stimulating and *more* exciting than Disneyland to get that dopamine rush!

When our environment is overstimulating, we experience frequent dopamine releases. In order to maintain homeostasis, the brain adapts and adjusts by reducing its natural dopamine production. As a result, the very experiences that once elicited pleasure and excitement become necessary just to uphold a baseline level of contentment. We now *need* them to maintain a stable baseline. But they no longer make us feel *good*. We need them simply to feel *normal*.

This is how tolerance works. Our brain gradually learns to recalibrate its baseline and we become desensitised to the smaller yet inherently healthier dopamine rewards. Simple day-to-day pleasures like a leisurely stroll in the park or reminiscing over cherished photos lose their effect if we are used to getting strong and frequent dopamine releases.

This brings us to the subject of addiction. I want to reiterate that this book is *not* a self-help addiction recovery guide. Addiction is a serious mental health concern that requires intensive and specific treatment. Abstaining from addictive substances like drugs and alcohol is *not* recommended unless done under the guidance of a medical professional, as the detox and withdrawal symptoms can be dangerous and even life-threatening. This book is instead for those of us who find ourselves on a slippery slope, spending that extra half hour before bed scrolling through TikTok aimlessly, or continually checking our phones for notifications, or telling ourselves 'I'll just watch one more episode . . .' It is important to understand how that initial dopamine pull can turn into an addiction, or at least a problem, without us even realising.

Professor Mark Griffiths from Nottingham Trent University in the UK developed what's known as the 'components model' of addiction.[1] This model outlines six key elements that apply to any kind of addiction:

1. **Salience:** This refers to an activity becoming a very important part of a person's life. When this occurs, it often dominates their thoughts, feelings and behaviours. If they aren't engaged with that activity, they will likely often think about when they next can.

2. **Mood modification:** This refers to how an activity alters a person's mood and it may vary each time they engage with the particular substance or activity. It might feel like it produces a 'buzz' or a 'high', or it might numb or help them feel a sense of escape from real life. Mood-modifying experiences are common for both substances and behavioural addictions. People use these activities as a way to both cope with emotions and experience alternate emotions.

3. **Tolerance:** With repetition and over time, a tolerance is developed. What once created a dopamine 'high' no longer does or does so to a lesser degree. Due to the tolerance, a person needs 'more' to achieve the desired feeling. This could include

consuming more alcohol, spending more money, gambling more, or getting more likes on Instagram.

4. **Withdrawal:** The brain adjusts its baseline, such that some form of stimulation is required to maintain homeostasis. This means that we shift from a state where something produces a dopamine release, to a situation where *not* having that substance or activity disrupts homeostasis and we enter withdrawal. Withdrawals are unpleasant feelings or physical reactions. This is often when people may feel cravings and urges. Suddenly, we *need* that substance or behaviour to feel normal, stable, or just okay. You'll know if you're having a craving because the sensations are intense and overwhelming. Our baseline dips below its new normal, and we feel uncomfortable psychological and/or physical symptoms.

5. **Conflict:** This can refer to both the conflict a person may experience with those around them, as well as internal conflict. Choosing short-term, dopamine-driven pleasure and relief can result in longer-term problems. A person may be aware that they want to reduce a particular activity but feel unable to do so and experience a sense of loss of control.

6. **Relapse:** There's a tendency for people wanting to stop or change their relationship with a chosen activity to experience relapses. These are essentially periods where someone reverts to a previous pattern from which they were aiming to abstain. These relapses can occur very quickly and a person can return to their previous level of use quite rapidly.

Depending on the type and severity of an addiction, the cravings and withdrawal symptoms can range from mildly uncomfortable states all the way through to highly distressing and all-consuming experiences.

For people who develop a diagnosable addiction, there are many factors at play, including biology and genetics, the environment, and the nature of the substance or behaviour. Addiction has historically been viewed through a black-and-white lens for the purposes of diagnosis and treatment. You either have it, or you don't. In some sense, this is true. In order for a health professional to diagnose an addiction, there are certain key features and symptoms that a person presents with, and to a certain degree of severity. As mentioned above, a tolerance builds up over time, the behaviour feels out of control, there are withdrawals when trying to cut down or stop, and there are often relapses. But the black-and-white threshold exists only for the purposes of diagnosis and treatment. In reality, addiction exists on a spectrum.

We can experience harm without being addicted to something. For example, a person may have a tendency to gamble a lot, which means their finances are affected, they have less money to go out and socialise, and the time they spend gambling instead of with their family causes tension. They might not have a gambling addiction per se, but they are experiencing what we call 'gambling-related harm' in other areas of their life. The same applies to alcohol. Someone might go out and drink heavily and wake up with a hangover. They might not be addicted but they end up getting to work late and performing poorly the next day. Similarly, spending too much time on social media might not indicate an addiction, but it may result in us being less present and more distracted when catching up with friends, and reduce the quality of our interactions with others. I may not be addicted to sugar or those tasty doughnuts from the café beneath my clinic, but eating too many of them will be bad for my health.

Most of us have some sort of problematic relationship to something in our lives, whether that's coffee, alcohol, social media, video games, or our mobile phones. For me, it's my emails. I check them way too often! On the train. Waiting in line for a coffee. In the office. I'm guilty of having woken up in the middle of the night and checking them then too! Research indicates that problematic and unhealthy social media use is on the rise, with younger people (particularly young females) most at risk.[2] Problematic mobile phone use in Australia has been increasing since 2005, again with young people aged eighteen to twenty-five and females experiencing the most difficulties.[3] The rise of problematic video game use led the World Health Organization to add it to its list of 'diseases' in 2018.[4] Of all the problematic online behaviours, online gambling and online shopping have been found to be the most common, again affecting younger people more.[5] It's safe to say that problematic use of technology is rampant, and unfortunately there are no signs it's slowing down anytime soon.

If we try to abstain from checking our phone, looking at social media, drinking coffee or alcohol, eating sugary foods or anything that we do regularly that gives us a dopamine hit, we'll likely experience discomfort, cravings and withdrawals. This is because our brain enters into a state of dopamine deficiency. While previously, a sugary treat or game on our phone might have given us a dopamine boost, repeated use results in our brain adapting and resetting its baseline. We may start to feel irritable, anxious, restless and frustrated without that activity or substance. In our world of constant stimulation, most of us are faced with this problem. While we might not all have a diagnosable 'addiction', we struggle to navigate the pull of dopamine hits, the tolerance that builds, and how to make changes so that we still have some pleasure and joy in life, while also engaging in the things that are meaningful to us.

Withdrawals and cravings drag us away from doing what's important or meaningful. There have been times where I've had something important to do. Perhaps a research paper to write with a deadline for submission. This has been important to my work and my career, and yet despite this, throughout the writing process I find myself continuously checking my LinkedIn, Instagram, Facebook and possibly even my bank statement on my phone! This urge to constantly check interrupts the focused work I'm trying to do. In fact, while writing this paragraph I'm feeling the urge to check my phone and social media accounts! But I'm going to consciously resist the temptation. It'll all still be there in an hour's time.

You may find that you're at a family barbecue, or your kid's soccer game, and yet the desire to check your phone is irresistible. You may love watching sports with your friends but the desire to check the odds is even stronger. In this way, the things which once felt good are now needed for us to feel okay.

I hope you now have some understanding of how dopamine works, but also how we can easily become unstuck in an over-stimulating world. In the next chapter, we'll look more closely at specific behaviours such as online shopping, pornography, drug use, social media use and video games to understand how these interact with our dopamine levels.

LET'S REFLECT

- Are there activities or substances in your life that you feel you have an unhealthy relationship with? Something that you do too often, too automatically and too compulsively?

- Have you ever tried to reduce or stop this behaviour, but experienced cravings and a desire to go back to it?

- Have you ever had the urge or craving to do something? What did it feel like in your body? Were you able to ride out that urge and not succumb, or did the dopamine pull you in?

Chapter 4

Chasing pleasure, escaping pain

In this chapter, we'll take a closer look at how dopamine is linked to various behaviours, activities and substances. We've talked about how it triggers pleasurable feelings, but what happens when we chase those feelings constantly? Dopamine plays a crucial role in the development of problematic relationships with substances like drugs and alcohol, as well as behaviours like gambling, gaming and sex.

Addiction doesn't discriminate. It doesn't matter where you're from, what job you have, how much money you earn, your nationality or your age; we can all be vulnerable. And no matter who we are, there are risks and experiences that can seem unrelated to our day-to-day lives. They appear so removed that it's almost impossible to imagine them affecting us. This is what psychologists refer to as the 'optimism bias' or, as I like to call it, the 'it-won't-happen-to-me' effect. The optimism bias is our tendency to overestimate the likelihood of positive outcomes, while underestimating the likelihood of negative ones. This then leads us to expect that things will work out well, despite our rational understanding that problems, challenges and setbacks are inevitable. It's one of the reasons why people indulge in risky behaviours, such as taking drugs at parties, speeding while driving, or going to the beach without putting on sunscreen.

Let's find out more about how the reward pathways in our brain are activated by certain substances and behaviours.

Alcohol

I'm that friend at a party who'll make you either not want to drink, or not be my friend. Perhaps my work with addiction has made me more vigilant. I'm known to cite the World Health Organization guidelines, ask people how many drinks they've had, and implore them to drink plenty of water. And I frequently post online about the misconceptions and harms attached to alcohol. That's not to say I don't enjoy a drink. I'm quite fond of a smoky whisky or an amaretto sour. But knowledge is power, and with the right knowledge we can make good decisions.

Alcohol addiction is common. Unfortunately, the messaging surrounding alcohol has led many people to believe it's safer than it really is. A glass of red wine a day was previously said to be good for your health. It was suggested that alcohol protected you against certain health conditions. The reality is quite different. The risks and harms associated with drinking have been systematically evaluated over many years and are very well documented. In 2023, the World Health Organization stated that there is *no safe amount of alcohol* that we should be drinking.[1]

In Australia, the latest government guidelines state that in order to reduce the risk of harm from alcohol-related disease or injury, healthy adults should drink no more than ten standard drinks a week and no more than four standard drinks on any one day.[2] Keep in mind one full-strength can of beer equates to around 1.4 standard drinks while the average glass of wine served in a restaurant is around 1.6 standard drinks. *One standard drink does not equal one alcoholic beverage.* More than a quarter of all adults exceeded the guidelines in 2022, and more than a third of young adults (aged eighteen to twenty-four) did the same, with males more likely to drink higher quantities.[3]

Alcohol is a toxic, dependence-producing substance. Decades ago, it was classified as a Group 1 carcinogen by the International Agency for Research on Cancer (IARC). This means that it's in the *highest risk group* for developing cancer, and in the same category as asbestos, radiation and tobacco! It's known to cause at least seven types of cancer, including breast cancer and bowel cancer.[4] Ethanol, which is the ingredient that affects our health and makes us feel drunk, is the chief culprit. Any beverage which contains ethanol, regardless of its cost or quality, increases our risk of developing cancer.[5] It doesn't matter whether it's a simple 'goon bag' or the most extravagant French champagne; *both* present a risk to our health.

Alcohol also contributes to high blood pressure, heart disease, stroke and liver disease.[6] But beyond these serious health conditions, we risk developing a dependence. Ethanol is a 'neurotropic substance', which means it crosses into the blood–brain barrier and inhibits the functions of our central nervous system. It affects the brain in a couple of significant ways. Firstly, it enhances the effects of a neurotransmitter called GABA. Through GABA, alcohol slows down the brain's functioning and neural activity, depressing the central nervous system.[7] This is why when we drink a lot, our speech becomes slurred, we're unsteady in our movements, we can't react quickly and our ability to think rationally is compromised.

Secondly, alcohol affects dopamine. Consuming even small amounts can increase the release of dopamine (specifically, within the nucleus accumbens, a part of the brain which plays a primary role in our experiences of pleasure, reward and addiction).[8] The activation of the dopamine reward system is important because it facilitates the salience and pleasure we derive from alcohol. But it can also contribute to compulsively seeking out the rewarding effects and push people to drink even more in an attempt to increase those feel-good sensations. A depressed nervous system and the release of dopamine can easily send us down the path to harm and dependence.

Drugs

This is an enormously complex topic so I'll simply provide an overview of a few types of drugs and the role dopamine plays. If you're interested to find out more, check out the Alcohol and Drug Foundation website (adf.org.au/drug-facts/).

Stimulants such as cocaine or amphetamines excite the brain's dopamine system like nothing else. No natural behaviours or activities have the same impact on the neural circuit of desire, not even food or sex! These drugs artificially stimulate the release of dopamine and disrupt our natural cycles, which is what makes them so dangerous and so addictive.

Not all drugs trigger the same level of dopamine release, but generally speaking, those that trigger a greater release are more addictive. They work by inhibiting the re-uptake of dopamine in the brain. In simple terms, this means the drug blocks dopamine from being absorbed and it remains in the 'synaptic gaps' of the brain for longer. For example, cocaine blocks the uptake of the dopamine from the synapses, resulting in dopamine accumulating and producing an amplified and intensely pleasurable high. But over time, we run the risk of developing a tolerance and becoming addicted.

With opioids such as heroin, morphine and prescription painkillers, dopamine works differently. Opioids bind to the opioid receptors in the brain and stop electrical pulses and messages from neurotransmitters travelling through the nerve cells, or neurons. This is helpful when it comes to pain. Imagine that you have chronic back pain: your back muscles are sending constant messages to your brain telling it that it hurts! Opioids block the messages being received which lessens the pain. Dopamine is also released and produces a feeling of euphoria – along with the pain relief. This reinforces the positive effect of the opioid, meaning that someone may be more likely to use it again in the future. In addition, the dopamine acts within a part of the brain called the amygdala, which produces the effect of relieving anxiety and stress. In this sense, opioids do a whole range of things that are experienced as pleasant – they stop or reduce pain, provide a sense of euphoria, and reduce stress and anxiety. This is why we have seen epidemics of opioid addictions, particularly in the US, where unregulated prescription of medications such as fentanyl and oxycodone have led to serious problematic use.

Social media

Unfortunately, when it comes to social media, we start off at a disadvantage. This is because social media apps have been designed to be reinforcing, or as they say in the tech world, 'sticky'. The developers making these platforms have a deep understanding of people, human psychology and how the brain works. If their primary motivator is profit – which in most cases it is – then they need to keep people on and using their apps as much as possible. Consequently, the apps need to be as attractive and as irresistible as can be, enticing us to spend more time on them, and creating a persistent urge to return.

So, what exactly is it about social media that makes it so hard for us to resist? There are various strategies being employed, but one in particular is the same psychological principle that keeps people playing the pokies. It's called 'intermittent reinforcement' and the techniques used are how psychologists teach parents to shape a child's behaviour, and even how dog trainers train dogs!

Intermittent reinforcement works by giving a reward at *irregular* intervals. If you want to teach a dog to sit, you say 'sit' and every time it does so, you give it a treat (a reward). Once that behaviour has been established, we maintain it by creating the *anticipation* of a reward. If the reward continued to come every single time the dog sat, that would be nice, but it would also become boring. If the dog didn't feel like a treat at that time and was asked to 'sit', why bother obeying the command? Instead, we reinforce the behaviour in intermittent or irregular intervals. This makes receiving a reward unpredictable, and it teaches the dog that even though the reward may not have come *this* time, it will eventually come. The dog is thus driven to continue obeying the command knowing that a reward will come at some point; it's just not sure exactly when.

In a similar way, social media has been explicitly designed to be *unpredictable*. When you pull down and refresh your newsfeed, or re-open an app, you'll see something new. You don't know what those new posts will be. They might be very interesting. They might be mildly interesting. They might be as boring as they get. But we keep checking and refreshing because we know sometimes we see something really good. The same goes for social media interactions. When we post a photo, we don't know how many likes it'll get, who'll like it, or when the likes will appear. As a result, we keep on checking.

Poker machines work in very similar ways and are known to be highly addictive, hence the rules and regulations we have around them (although some might argue we need even more regulations, but that's another conversation!). If pokie players knew the outcome every time they pressed the button, there would be less of a pull and a desire to play. It's the unpredictability and anticipation that creates the desire to keep swiping, checking or playing.

Now this is where the developers and software engineers get really crafty. Rumour has it that apps such as Instagram will sometimes 'withhold' likes when you post a photo.[9] When you make your post, you may initially be disappointed at the number of reactions. It's been suggested that the app will withhold likes and then release them later on, usually in a larger burst all at once. Why? Because this triggers those dopamine pathways. Remember, dopamine acts in the *pursuit* of a desired outcome as well as its attainment. If 'lots of likes' is the desired outcome for posting on Instagram, the app will withhold them to keep you in a state of pursuit and anticipation. When you eventually do get that big release of likes, the dopamine hit is higher because it comes at once. Our brain is primed to respond to the sudden influx of positive social approval. While it's not as intense as a line of cocaine or winning the jackpot on the pokies, the positive social appraisal and interactions result in a release of dopamine, reinforcing the behaviour (in this case social media activity).

Other strategies that make social media intensely appealing include socially rewarding stimuli – smiling faces, love hearts and likes, for example, are all forms of approval. Think about how different it feels for someone to 'heart' a photo compared to something unemotional like giving it a 'tick' or a 'thumbs up'. These types of social stimuli are rewarding and activate those dopaminergic pathways. You may also have noticed that over time, the criteria to receive a notification has changed, and there are more and more ways to engage socially – we can post photos, videos, reels, stories, polls, question boxes and so much more. Infinite ways to pursue social reinforcement, and infinite reasons to open the app and keep checking. This is dopamine and social media at work.

Consider also how easy and effortless it is to use social media. It doesn't require much cognitive capacity at all. It was made even easier when Facebook ditched its 'click to see more' function when you reached the bottom of your feed. You can now automatically keep scrolling forever. Facebook removed the requirement for you to make an 'active' decision to keep scrolling by pressing the button.

This button was a disruption to the trance-like state we get into when scrolling. So it was removed.

In the clinic recently, I've also noticed an increase in unhealthy relationships with dating apps. These apps are designed to facilitate connections, but they can also result in addictive-like behaviours, as people find themselves constantly swiping, matching and seeking validation and social connection. Being presented with one profile at a time is just like refreshing your feed or notifications on social media. Swipe yes or no, and you'll then be presented with someone new. The allure of gratification and the dopamine rush from receiving matches can lead to compulsive usage.

Gambling

Here in Australia, we are a nation of gamblers. Gambling is a major public policy issue, with Aussies losing approximately $25 billion each year on legal forms of gambling. This equates to us having the largest per capita losses in the world.[10]

When people gamble and win money, their brain releases dopamine, generating feelings of pleasure and reward. This dopamine surge not only reinforces the desire to continue gambling, but also intensifies the motivation to chase that euphoric sensation. The allure extends beyond the monetary gain, as it also compels people to seek out the thrill and excitement of *potentially* winning.

Have you ever heard of a 'near miss'? This is the experience of *almost* winning. For example, if on a poker machine a 'win' is considered getting three cherries in a row, then a 'near miss' would be getting two cherries and a lemon. It's the idea that I *nearly* won, and I *just* missed out. Now, let's remember, dopamine is involved in the *pursuit* of a reward and not only its attainment. This means that it's triggered during a near miss experience and encourages further gambling. When a person experiences a near miss, the brain reacts in a similar way to when a win occurs. This is because the brain perceives the near miss as a *partial* reward, tricking the gambler into thinking that they are progressing towards a win, even though they didn't technically win anything. In fact, they lost money! What we don't take into account is the fact that each spin on a poker machine is *independent* and *random*. It has nothing to do with the spin that came before it, or the one that comes after it. The notion that I have a near miss so I must be *close* to winning is simply false. Despite this, the dopamine activation that occurred during the near miss compels us to keep playing and creates a sense of excitement and anticipation in search of that next win.

As with social media, gambling functions on an 'intermittent reinforcement' schedule. We don't know when or if a win is going to come, but we do know that wins happen because we've either experienced them ourselves, or know of other people who have. Statistically, a win *is* possible, but it's the *unpredictability* that triggers the dopamine response in the brain. The anticipation of the reward becomes enticing. Winning big when someone first starts gambling is a risk factor for later problematic gambling.[11] Dopamine plays a role in this process, as a person continues to chase that initial 'high' produced from the big win.

People addicted to gambling and drugs describe the euphoric feelings in much the same way. The more money gamblers lose, the more they tend to persevere to win it back. This is known as 'chasing losses'. When someone loses money, they feel compelled to keep playing in the hope of recouping their loss. Unfortunately, this usually results in *more* losses. The experience shifts from that of 'liking' to 'wanting'.

The pokies are more addictive than lotteries or scratchies. Why is this? Let's think back to social media. One of the features that keeps us hooked is how simple it is to keep playing. We can do it very rapidly, with fast continuous play and little time between bets and outcomes. Along with the rapid pace, there is immediate feedback (that is, the outcome of whether someone wins or loses), while the lights, sounds, vibrant graphics and reinforcing sensory stimulation makes it easy for someone to become engrossed and continue playing.

Pornography

In recent years, we have seen an increasing acceptance that not only drugs, alcohol and social media lead to addiction. Pornography also triggers the release of dopamine and its overuse can become problematic. Behaviours become addictive when they are repeatedly reinforced in the reward, motivation and memory pathways of the brain. There is currently no universally accepted diagnostic criteria for a pornography addiction, but there's no doubt that it can, for some people, result in serious and harmful consequences. Let's talk about why.

During sexual activity, whether it's in real life, or while watching pornography, dopamine is released from a part of the brain called the ventral segmental area (VTA).[12] The release communicates to other parts of the brain about how well our needs are being satisfied and reinforces the motivation to keep seeking out these feelings, resulting in the desire to continue the activity. This is the same for other naturally occurring behaviours, from eating food to exercising, all of which activate rewards systems that are necessary for survival.[13]

However, over time, in the case of pornography, the brain builds up a tolerance to the excess dopamine and requires either more or different content to produce the same level of pleasure. In a study of some 6500 people in Poland, the adverse effects of pornography included the need for longer and more sexual stimulation to reach orgasm, as well as an overall decrease in sexual satisfaction.[14] It seems to be greater for those who start using pornography in their teenage years. It has been speculated that a tolerance to pornography can lead to seeking out more intense and aggressive varieties, with further links to real-life sexual violence. Researchers have also discovered links between pornography and aggression, whereby watching violent material over time predicts an increase in the odds of self-reported sexually aggressive behaviour. This may however be due to the fact that more aggressive males prefer to watch violent pornography.[15] Further research and exploration is required before any firm conclusions can be drawn.

Shopping

When we think about buying something, our brain begins to release dopamine. Anticipation and desire start to build if it's something we want. As we browse in a store or search online, our brain continues to release dopamine as we see different items of interest and consider buying them. When we eventually do decide to buy, our dopamine levels spike again, resulting in a rewarding experience, culminating in feelings of fulfilment and satisfaction. This surge in dopamine reinforces the behaviour and establishes a connection between the act of shopping and the pleasurable experience it brings.

What the brain loves even more than a reward is an *unexpected* reward. Imagine you're browsing online on a Sunday afternoon. You stumble upon an ad for a great outdoor barbecue. You've been thinking lately about how nice it would be to have some friends over in summer, and how useful a barbecue would be. Next thing you know, you've got multiple browser tabs open with different barbecues, and are googling reviews of different makes and models. You decide to buy, place the order and feel great. You weren't looking to buy a barbecue that Sunday afternoon, but the reward of unexpectedly finding and buying one is even greater than it would be if you'd planned it.

There are other factors at play that can make the allure of shopping a hyper-stimulating experience.[16] The faster a process can be completed, the more addictive it can be. Conversely, the longer something takes, the less addictive it usually is. Imagine if a poker machine took two minutes to spin its reels. People would lose interest pretty quickly. With seamless online payment systems, shopping has become effortlessly accessible. Couple this with sophisticated marketing techniques that hook consumers into wanting to buy a product and you have a potent combination. Through the advertising of seemingly good sales and making products seem scarce, consumers feel the urge to splurge. Online shopping, easy payments and fast deliveries all contribute to the dopamine experience. And as technology has developed, the protective delay between impulse and purchase has decreased.

Online shopping retailers have also caught on to the value of 'gamification' features within the shopping experience. Gamification is where elements of games, such as scoring points and competitions with others, are used as an online marketing technique to keep people engaged with a product or service. The online shopping platform, Temu, is well known for incentivising its customers through game-like features. The blend of shopping and games creates an enticing user experience, with bonuses and coupons that mimic the rewards you might get from a video game. Other brands around the world are getting onboard with gamification, incorporating 'spin to win' features and quizzes on their sites. Starbucks rewards customers with stars for purchases and offers challenges to earn bonus points. Lacoste has a crocodile hunt game in which shoppers can unlock prizes in its virtual store.

A client by the name of Sally, whom I worked with for some time, struggled to manage the impulse and excitement of online shopping. She particularly found herself drawn to shopping retailers like Temu and Shein, where the options for purchases are endless. The combination of low prices, countdown timers and gamification features made it easy to become hooked. The cheap prices give consumers a dopamine hit and keep them coming back for more. While I wouldn't necessarily say that Sally was 'addicted' to online shopping, she was spending more than her disposable income allowed.

Video games

Just like with gambling, social media and pornography, it's easy to become addicted to gaming. There are many reasons why. Winning in-game trophies, getting recognised on leader boards, or achieving rewards that will help with the next stage of the game all result in a sense of accomplishment, which activates the brain's reward pathways. The positive reinforcement encourages a person to continue playing and engaging. As a player progresses, dopamine creates a sense of achievement and motivates further play. The rewards are often instantly received, and this quick feedback loop enhances the overall experience and provides immediate gratification. Add into the mix immersive and engaging narratives and well-developed characters and a video game can provide a rich source of escapism from real life. You can also play at home at any time of the day or night.

The same goes for mobile app games which, like other forms of gaming, are specifically designed to keep you playing. The escapism created means that our sense of time often feels warped when we're playing. We become so immersed that our brain doesn't keep track of actual time and instead measures it according to the pleasure we're experiencing playing the game. Reaching the next level becomes the next marker of a passage of time. The bright colours and exciting sounds grab our attention, while also creating a classical 'conditional loop' that is reinforced by dopamine.

When people who play games on their smartphones were asked what makes a game more or less 'addictive', three features stood out.[17] Firstly, games which are 'challenging' are more likely to hook people in with a desire to progress and accumulate in-game rewards. Secondly, the *social* element of being able to play with family and friends – interacting as a group, working together strategically, building virtual friendships and achieving social recognition – is highly rewarding. And the third enticing factor gamers cited was the graphics and animations.[18]

Since dopamine is involved not just in the experience of pleasure, but also the *pursuit* of pleasure, each time we pick up our phone, we experience the urge to want to open an app and play a game, activating the same pathways in the brain responsible for craving in substance dependence.[19]

For most people, playing video games is a fun and entertaining pastime, but there is a minority who experience serious problems. Brain imaging studies have showed that gaming shares the same neurobiological activations as other addictions.[20] Dopamine and the reward pathways are activated. Furthermore, people who become addicted to video games show changes to their 'grey matter' volume in the dopaminergic system.[21] This is thought to contribute to impaired self-control, which further perpetuates the problem.

Professor Warburton from Macquarie University has spent a lot of time researching gaming and screen addictions and is very concerned about the implications of problematic screen use. He says, 'Video gaming is currently the only type of problematic screen use with a clinical diagnosis attached, but the more research we do, the clearer it is that all types of screen use are the same. It doesn't matter whether it's online or offline gaming, internet browsing, social media, or smartphones, all these platforms are built on an addiction model, and they can all cause the same problems.'[22]

This doesn't mean that everyone who plays video games will develop an addiction. However, for people who do have a tendency towards addictive behaviours, or who might have an underlying vulnerability such as mental health stressors, there may be a greater chance of getting hooked. Caveat gamer!

What else?

As time has gone on, we've found more and more behaviours with which people can develop problematic relationships. One of the leading researchers in this field who you might remember from Chapter 3, Professor Mark Griffiths, has done a lot of work understanding which behaviours we can become 'addicted' to. He's looked at excessive internet usage, overworking, binge-watching content,[23] taking millions of selfies[24] and even studying[25] too much. Professor Warburton even warns us about an addiction to 'reeling' (watching short videos you see on platforms like TikTok, Instagram and YouTube Shorts). Reels are designed to give us lots of little bursts of dopamine, without ever getting the feeling of being satisfied. Because of AI algorithms, we keep getting shown more of what we like, so we keep scrolling to the next reel in search of something even more entertaining. Just like other types of interactive screen platforms, reels are designed to be 'sticky'.[26] Even dating apps can be addictive and manipulative, with some people claiming that they form an unhealthy relationship with the app, swiping constantly.[27]

There's still very limited research to support classing these behaviours as 'addictions', and more needs to be done to understand the neurological underpinnings and the ways in which the activation of dopamine is involved.

LET'S REFLECT

- How does this chapter make you reflect on your own relationship with pleasure-seeking activities or substances? Are there any behaviours or habits that particularly resonate?

- How might understanding the role of dopamine and the reward pathways in the brain influence how you engage with these substances and behaviours?

- Armed with this knowledge about dopamine and rewards, are there any changes you'd make?

Chapter 5

What other neurotransmitters are important?

The focus of this book is dopamine, dopamine-driven behaviours and reshaping our life to achieve a healthy balance between pleasure and purpose. However, dopamine doesn't act alone: other neurotransmitters such as endorphins, serotonin, oxytocin and noradrenaline are also crucial to our wellbeing. The 'good' feelings these neurotransmitters produce vary. For example, dopamine gives us a pleasant sensation with a sense of desire and can create a form of happiness. Endorphins mask pain and allow us to push through difficult situations. Serotonin elevates our mood and relieves anxiety. Oxytocin helps us feel attached, connected and bonded to other people while noradrenaline helps prepare our body for action. In the sections below, we'll examine the role of these four key neurotransmitters, how they work, what they're responsible for, and why they're so important.

Endorphins

Endorphins, along with dopamine, play a crucial role in our survival. Their job is to mask pain for a short amount of time. For example, if you were involved in an accident, endorphins kick in to temper the immediate pain, allowing your body to take action and escape a potentially dangerous situation.

Endorphins are also produced during activities such as exercise, causing a wave of pleasure.[1] They can relieve stress and put people in a good mood. Have you ever heard of the 'runner's high', or people becoming 'addicted' to jogging? Well, we have endorphins to thank for that. A 'runner's high' is experienced when you push yourself to the point of distress. When this happens, the body releases endorphins which allow you to continue to run despite the pain and stress on the body.

Endorphins share structurally similar properties to opioids such as morphine, heroin, codeine, oxycodone and fentanyl, and yield similar effects. These drugs block pain signals in the brain. They can also reduce stress and put you in a euphoric or drowsy state of mind. It's no wonder they're so addictive! The opioid epidemic is a major public health concern worldwide, though it's particularly bad in the US. In 2021, the fentanyl category of opioids accounted for 67,325 preventable deaths, a 26 per cent increase from the previous year.[2] Fentanyl is prescribed for chronic pain, severe cancer pain, nerve damage, back injuries and surgeries. In Australia, it is a Schedule 8 drug, along with morphine and oxycodone, meaning that there are very tight restrictions on doctors and pharmacists prescribing these drugs due to their high potential for addiction. And for good reason. Fentanyl is about *eighty to a hundred times* stronger than morphine.[3] Like all opioids, fentanyl is a respiratory depressant, meaning that it can interfere with a person's ability to breathe, easily resulting in accidental overdoses.

Oxytocin

Remember the warm, fuzzy 'feel good' sensations you sometimes get on a first date? Or how about the inner warmth you feel when you hug a loved one? Or maybe the joy of cuddling a newborn baby? That is oxytocin working its magic. Oxytocin is a hormone that acts as a neurotransmitter in the brain. And one of its primary functions is to help us to bond, connect and feel attached to other people.

Humans are inherently 'social animals' and it is the need for social attachment that drives a lot of our behaviours. Oxytocin helps us feel bonded. It regulates our emotions and encourages us to be sociable. It also helps us feel trust and empathy. It's why we feel good when we're around people we care about. Oxytocin is also important for survival. It's released during sex and for this reason it's commonly referred to as the 'love hormone'.

From an evolutionary perspective, oxytocin plays an important role in attachment, particularly as a critical maternal hormone, allowing parents to bond with their children. In mammals, oxytocin is associated with childbirth as well as lactation.[4] In the hours before the onset of labour, women experience a rise in oestrogen which triggers a significant increase in the production of oxytocin.[5]

A 2005 research paper showed that oxytocin was also key to trust among humans.[6] The release of oxytocin resulted in a substantial increase in trust, and subsequently increased the benefits received from social interactions.[7] Specifically, it affects a person's willingness to accept social risks. Giving people oxytocin even increased their generosity towards others.[8]

We also need to consider the role oxytocin plays in loneliness. Loneliness is a growing problem, something the World Health Organization has identified as a global public health priority.[9] We all feel lonely from time to time – it's a normal and expected human experience – but *chronic* loneliness, where we feel a lack of connection to other people and a lack of satisfying social relationships, has major health and mental health consequences. Pre-COVID-19, increasing rates of loneliness were already worrying. Between 2001 and 2021, one in five Australians agreed with the statement 'I often feel very lonely'.[10] Loneliness is associated with a 26 per cent increase in premature death.[11] It increases our risk of high blood pressure, faster progression of Alzheimer's disease, worse cognitive functioning, decreased immunity, and increased inflammatory conditions, depression and anxiety.[12] Low oxytocin levels contribute to feelings of loneliness, and that may mean a greater susceptibility to negative health outcomes.[13]

So how can we boost our mood and enhance our feeling of connectedness to others? By engaging in meaningful connection. Physical touch triggers the release of oxytocin. The simple act of giving someone a hug hello or goodbye, putting an arm around their shoulder or holding their hand can feel good for both the giver and receiver. Connecting physically and emotionally with those we love and care for produces oxytocin, helping us feel good and reducing stress. So don't be a stranger – go and give your loved ones a hug!

Serotonin

When most people think of serotonin, they think of the brain. However, only about 10 per cent of our serotonin is produced there, while the rest is found in the intestines and the cells lining the gastrointestinal tract! When our serotonin levels are balanced, we feel focused, calm and stable. Increased serotonin within the central nervous system elevates our mood and reduces anxiety.[14]

The link between lower levels of serotonin and depression was first suggested in the 1960s.[15] The theory is now well known, with surveys showing that approximately 80 per cent of Australians believe that depression is caused by a 'chemical imbalance',[16] with many doctors agreeing.[17] However, it is in fact still just that – a theory. The evidence is inconsistent and there's no consensus that depression is caused by a lower concentration of serotonin.[18]

This is of course not to suggest that you go and throw out your antidepressant medications! Drugs such as Selective Serotonin Reuptake Inhibitors (SSRIs) are very effective for many people. The takeaway is that the brain is incredibly complex and there is still so much that scientists and doctors don't know.

Norepinephrine (noradrenaline)

Norepinephrine, also known as noradrenaline (which is how I will refer to it here) is both a hormone and a neurotransmitter. Its main function is to mobilise the brain and body for action. Noradrenaline increases our general alertness and attentiveness and enables us to move. It also promotes vigilance, focuses our attention and enhances memory formations.

Low levels of noradrenaline are released during sleep because this isn't when we need to be alert or ready for action. When we wake in the morning, the brain sends a small burst of noradrenaline into the bloodstream to help us get going. If when exercising you get an extra boost of energy, that's noradrenaline helping you power through. The highest levels are released during stressful or dangerous situations.

There is a difference however between noradrenaline and adrenaline. Noradrenaline is *continuously* released at low levels, while adrenaline is *only* released during times of stress. When your brain senses danger, it triggers a flood of adrenaline to activate the 'fight or flight' response. This helps us to prepare for combat or enables us to flee. Adrenaline increases our heart rate and blood pressure, triggers the release of glucose from our energy stores, and increases blood flow to our muscles but reduces it to our bladder and gastro-intestinal system.

During times of stress, we want our blood flow to go to our muscles to enable us to move, and our glucose stores to be released so that we have the energy to act. Less blood flow goes to the bladder and intestines because a stressful, adrenaline-fuelled situation is usually *not* the ideal time to need to use the bathroom. Thankfully our body knows that!

•

So how can we use *natural* strategies to boost our mood and well-being? There are three key activities that are good for our brain and body and potentially even better when combined. They are music, social connection and exercise.

Music activates both the brain's left and right hemispheres and assists with learning and memory. It improves mood, provides pleasure, can make a hard or painful task feel easier, helps us focus and increases our retention of information.[19] Research has shown that listening to music that you love makes your brain release more dopamine, giving you a pleasurable experience.[20]

But wait, there's more. Music doesn't just affect dopamine. It's also related to the release of endorphins (the neurotransmitter that helps mask pain). While listening to music triggers a low-level release of endorphins, actually singing, dancing or playing an instrument triggers an even higher release and an even more positive emotional state.[21] Now, let's build on that. What if we combine music, which releases dopamine and endorphins, with exercise, which also releases endorphins? Well, the outcome is extremely positive! When music is paired with exercise, it can motivate people to work harder and for longer than they even realise.[22] If you don't like listening to music while at the gym, or you aren't a fan of Zumba classes, next time you pull out the vacuum to clean, put on your favourite tune and sing along! It'll work wonders.

A colleague of mine, Kate Beever, an assistant professor at Berklee College of Music in the US, says there's no one style of music that works for everyone. It's an individual thing. If someone loves metal bands, go for it. It's just as effective as classical music or mellow folk songs.

> *When we listen to music we like, the experience is enhanced because we're paying more attention to it. We may even be singing along silently or tapping our toes. Even if we don't know the song, the structure of most popular music is familiar enough to our brains that the musical expectations set us up with anticipation. We have expectations about what's coming, and when those expectations are fulfilled, we feel a sense of relief, thanks to dopamine and serotonin being released. As far as we know, we can't become 'addicted' to music, and it doesn't lose its effectiveness – listening to music releases just enough dopamine to make us feel great.*

Let's now consider music and connection to others. Music in and of itself can help foster a sense of connection. After all, it's said to be the 'universal language'. We know that connecting with others releases oxytocin. Exercising with a friend while listening to music or attending a group exercise class adds a social element to the music and activity combination. Even better – why not go out dancing with friends? Combining music, movement and social connectedness could be the trifecta to feeling good!

While dopamine plays a key role in what we do and how we feel, it's important to remember the role of these other neurotransmitters. Dopamine doesn't act alone, but rather is part of a complex circuit in the brain. Harnessing their *combined* power is the natural way to improve mood and wellbeing.

LET'S REFLECT

- Can you recall any times in your life where adrenaline might have helped you?

- Are there ways that you can naturally engage in behaviours that promote healthy production of endorphins, serotonin and oxytocin?

- Reflecting on the role of these neurotransmitters and what activates them, are there ways in which you can naturally boost your levels of connection and happiness?

Chapter 6

The conflict of pleasure and purpose

Why do we need to know about dopamine and how to manage it? What's wrong with living a hedonic life? Why can't we just spend money, play video games, gamble, scroll social media and eat dough-nuts every day if it gives us pleasure and helps us feel good? There is nothing inherently wrong with living life in this way. However, it won't make us feel *fulfilled*. All we'll experience is a shallow, super-ficial happiness that is short lived. Think about playing a game on your phone. It's enjoyable at the time, but once we put the phone down and look up at our surroundings, we're confronted with life as it is. And that reality will never be free of all pain or struggle, but it is hopefully fulfilling and meaningful. If it's not, there's an even higher chance that we'll pick up our phone again.

Video games, mobile apps, gambling, drinking, online shopping and so much more, only give us a feeling of pleasure until our dopamine levels adjust. After that, these activities just feel 'normal'. Of course, we can't be eating, drinking, scrolling or gaming 24/7. We have to go to work, tend to responsibilities, pay the bills, cook dinner for the family, do the laundry, mow the lawn . . . the list goes on! If the way we feel 'good' in life derives from dopamine-driven behaviours, when we are not engaged in those activities, we're going to feel pretty uncomfortable. We'll probably feel sad, empty, unfulfilled or restless. We'll probably want to return to those dopamine-driven behaviours because they don't make us feel bad. At least not in the short term.

Herein lies the challenge of balancing pleasure and purpose. Don't get me wrong, I'm all for an Instagram scroll or an online purchase. I am not telling you to throw out the video games and dating apps. I'm actually a big advocate for their positive benefits. What I do want to help you do, instead, is find that sweet spot of equilibrium. I want to help you enjoy the things that feel good, while also living a life that is full, enriching and meaningful; one that's brimming with purpose and satisfaction. When you get to your eightieth birthday, I want you to be able to look back and reflect on your life, satisfied with the decisions you made and the way you chose to live.

Pleasure and purpose are not mutually exclusive. When we get the balance right, they work together to help us live an immensely rich life. Nor is the idea of balancing pleasurable activities and meaningful pursuits new. The ancient Greeks believed that there were two types of 'happiness' – a *hedonic* happiness, derived from the pursuit of pleasure or enjoyment and avoidance of pain; and a *eudaimonic* happiness, derived from the pursuit of meaningful endeavours. It is certainly possible to have one without the other, that is, a person can have a great deal of hedonic, dopamine-driven happiness, without the eudaimonic, fulfilling happiness. And the reverse is also possible.

Here are two examples. A client by the name of Garry works as a successful investment analyst. He earns a good income and lives a very comfortable lifestyle. He owns a large waterfront apartment in one of Sydney's most prestigious suburbs and several luxury cars. He has socialite status and is often invited to exclusive events. He dines at the finest restaurants, travels to exotic holiday destinations and attends high-end parties. Garry feels a lot of hedonic happiness and tells me that many of these things genuinely do make him happy. What he cannot find, however, is a feeling of purpose in life. Unbeknownst to his friends and family, he struggles with depression. He doesn't enjoy his work or find it meaningful, but it does allow him to continue living his hedonic lifestyle. Garry also often feels lonely and disconnected from others. He believes that many of his friendships are superficial, and that people only want to spend time with him because of his money.

Garry lacks meaningful pursuits. When I worked with him, we focused on discovering activities that were fulfilling and personally rewarding. He quickly realised that what was important to him were things that were 'bigger' than himself. He had a strong sense of connection to his culture. He came from a large Indian family and, growing up, this was a significant part of his identity. As he had progressed through life, he lost a sense of connection to his culture and tradition. He decided that he wanted to re-connect to his culture, and give back to the community that had provided him with such a strong sense of belonging when he was younger.

My recommendation was not to stop going on expensive holidays. Nor did I suggest he should downsize his waterfront apartment and move to a modest home in the suburbs. Rather, I recommended he try to find ways to *balance* his hedonic pursuits with meaningful and purposeful action.

There are only so many hours in a day. As you start to devote more time to meaningful pursuits, you'll have less time for hedonic happiness. Garry decided to attend more community events, while also donating money. This helped him feel more meaningfully connected to others, and his sense of loneliness gradually reduced. He naturally had less time for parties and extravagant social dinners. He would sometimes find himself faced with 'choice points' (more on this in Chapter 12). If there was a community event and a social event both on the same night, which would he choose to go to? There's no right or wrong here. What I did want to help Garry achieve is a sense of balance that felt right *for him*.

Now let's consider another client, Sandra. Sandra is a qualified social worker who's dedicated her life to helping underprivileged communities and people who've experienced trauma and poor mental health. Aside from her work, she also volunteers with various charities and is an active advocate for social justice and change. She finds her work incredibly meaningful, and making a positive difference in someone else's life, no matter how big or small, makes the hard work worthwhile. Sandra has a great sense of eudaimonic happiness in her life. She's fulfilled and feels as though she's able to make meaningful changes in society.

However, Sandra often feels tired, burnt out and flat from doing such emotionally draining work. Given the demands of her job and her commitment to various social causes, she has limited time for leisure activities. In fact, when she does have 'down time', she spends much of it doing things that are related to her work. She reads books about social justice or listens to podcasts of people who've experienced and overcome hardships. She lives a modest and frugal lifestyle and working long hours means she often sacrifices time with friends.

My focus was helping Sandra to achieve balance. We looked for activities purely for the purpose of pleasure that didn't serve a function other than being enjoyable. She really loved art and had taken art classes when she was young but didn't have the time for them anymore. So she bought a set of watercolour paints, some canvases and an easel, and set up a cosy corner in her home for painting and artistic expression. She also started watching some comedy on TV in the evenings that helped her unwind. As with Garry, naturally there were 'choice points' and not enough time to do everything. Sandra had to decide whether to pursue hedonic or eudaimonic pursuits; and balance was key in making these choices.

I am not here to tell you how much time you should spend indulging in those hedonic dopamine-driven pursuits versus how much time you should be giving back to the community or doing charity work. You need to find the balance that works for you, and there is no right or wrong.

Let me share a personal story with you. I travel to Broken Hill in rural New South Wales for work. It's a place that is incredibly special to me. The first time I went there, I fell in love. I fell in love with the people and their sense of community. I fell in love with the red, raw earth. And I fell in love with the feeling of peace and serenity in the middle of the outback. I've now travelled to Broken Hill and nearby Silverton four times, and hope to visit many more times in the future.

On a recent trip, I decided to dedicate some time to writing this book (after all, I had a deadline to meet!). But Wednesday night is karaoke night at the Palace Hotel. This is the hotel where *Priscilla Queen of the Desert* was partly filmed and is always a must on my Broken Hill itinerary. The first time I went to a karaoke night, I had a lot of fun! So naturally, my dopamine was firing, motivating me to return.

However, that Wednesday, I befriended a local Aboriginal family. One of the aunties, whom I had briefly met on a previous visit, gave me a big hug, introduced me to her mob and welcomed me with open arms and an open heart. She sat down and showed me photos from her childhood. There were photos of her siblings, some of whom were still alive and others who had passed. She showed me photos of her foster mother, who had also passed away recently, and photos of her foster siblings. She told me of fond memories as well as difficult times.

The family explained to me their connection to the land in the Broken Hill area, on Wilyakali country. They explained their totems and told me about the changing waterways in the region. The time I spent with this family was so special. Connection with others is very important to me. I was fully immersed in the knowledge and stories this family shared and felt connected to them through their openness. I felt a closer sense of connection to the land on which I was standing as well. And within me, I felt warmth, gratitude and fulfilment. Not a big 'laugh-out-loud' kind of happiness, but definitely a 'corners of the mouth turned up' happiness with a heart full of love.

Later that day, it was time for karaoke. Did I did go to the Palace Hotel and drink a ginger beer? I sure did! Did I sing along to the great Aussie classics that, without fail, always feature in regional town karaoke nights (Cold Chisel, John Farnham, some INXS and Lee Kernaghan's 'Boys from the Bush')? Yes! And did I post about it on Instagram and wait for people to like, comment and message me? Absolutely! I'm not immune to the dopamine hit of social approval.

When I got back to my hotel room that night, I had the memories of how fun the karaoke had been, but also how special my time with the local family was. They gave me different types of happiness. One could not possibly replace the other. It was a great day because I'd had balance. I'd had my fill of fun and pleasure, and I'd also had experiences that were meaningful and purposeful. I went to bed that night with a smile both on my face and in my heart.

Achieving a balance between pleasure and purpose is at the heart of living a rich life. I often say to clients, 'Nothing is a problem, unless it is a problem.' Problems do not lie with one specific behaviour. They lie in our *relationship* to that behaviour. They develop when that behaviour becomes automatic and excessive. There's no issue with me spending an evening binge-watching *Game of Thrones* – as long as I make the conscious choice to do that. It becomes problematic if every night after work, I get home, turn on the TV and watch for four hours. This behaviour is automatic and excessive and comes at a cost to other important things in my life.

We don't need to label one pursuit as 'right' and the other as 'wrong', we just need to find our own equilibrium. Just like Garry and Sandra, we all face choices where we must weigh up the drive for hedonic happiness against the pursuit of eudaimonic fulfilment. My day out in Broken Hill reminded me that *both* types of happiness are vital – and possible.

LET'S REFLECT

- What activities give you hedonic happiness?

- What pursuits give you eudaimonic happiness?

- Can you think of a time when you indulged in hedonic behaviours and felt a sense of emptiness or dissatisfaction?

- Have you ever felt torn between engaging in activities purely for pleasure versus those that align with your values and aspirations?

- How do you find the balance to be between hedonic and eudaimonic happiness in your life? Where is there room for change?

Chapter 7

Identifying your target behaviour

By this point in the book, I'm hoping that you've thought a bit about your own actions and behaviours. We discussed some different substances and activities that disrupt the production of our dopamine, as well as how these can get in the way of us living a life that is meaningful and purposeful.

I want to help you change your *relationship* with your problematic behaviours. Think of it as a power struggle. Right now, you might feel like something has power over you. You're doing it automatically and compulsively. Take for example those sugar-coated doughnuts from the café beneath my clinic. If I'm sitting at my desk each day thinking about when I can pop downstairs and buy one, then the doughnuts have power over me. I need to assert control over if and when I choose to have a doughnut, and not feel *compelled* to buy one, or preoccupied thinking about buying one. The same goes for people who binge-drink. They find that when they go to a party, they drink more than they'd really like to. They need to change their *relationship* with alcohol, so that there isn't the pull to drink to excess, and they regain control.

While we've talked a lot about the role that dopamine plays in shaping and driving our behaviours, there are other factors that influence how we might feel on a particular day. Our mood and emotions are big drivers of how much time we spend scrolling on TikTok or playing the pokies. Both positive and unpleasant emotions can influence us. Sometimes even boredom drives what we do. If we're hungry or tired because we haven't had enough food or sleep, we'll have fewer cognitive resources and less energy available to us to manage the day or assert power over certain behaviours. Yes, dopamine does play a role in driving us towards these behaviours, but other factors make us susceptible. We'll explore these in greater detail in Part 3 and examine how we can manage them.

I'd like you to choose a behaviour or substance you want to change your relationship with. Keep in mind the aim is not to quit it forever. You might want to make that your goal and, if so, good for you! But there are some behaviours that would be hard to quit completely (say, shopping, eating sugar and sex). For a period of time, we'll do our best to abstain, and then look at reintroducing the behaviour gradually and in a way that's more *balanced*. The goal here is to shift the balance so that the doughnuts, social media, alcohol or dating apps don't have the power in the relationship!

If you're still unsure which behaviour you want to change your relationship with, take a look at the list below. (Remember, if you think you have a serious addiction, don't do anything without consulting a medical professional first. Detoxing from certain substances and activities can be dangerous.)

- Alcohol
- Cigarettes
- Drugs
- Gambling

- Video gaming
- Shopping
- Sex
- Pornography
- Eating sugary foods/binge-eating
- Caffeine
- Social media
- Dating apps
- Exercising
- Binge-watching TV
- Working
- Thrill-seeking or risky behaviours

Once you've chosen the behaviour that you want to work on, keep it in mind as you read Parts 2 and 3. In Part 2, we take a break from dopamine and examine our values. We look at how knowing and understanding our values helps us to live a meaningful and fulfilled life, thereby reducing the automatic desire to escape it through our chosen problematic behaviour. And in Part 3, we start to make changes, taking specific action and practising specific skills.

Identifying a problematic behaviour is the starting point. You're already one step closer to living a life free of compulsions and dopamine-driven highs of instant gratification, and on your way to a truly fulfilling and meaningful existence. Congratulations!

LET'S REFLECT

- What obstacles do you think you'll face in trying to change your relationship with your chosen substance or behaviour?

- Reflect on the power dynamic between you and your chosen behaviour. In what ways does it currently exert control, and how would you like that dynamic to change?

- Reflect on a time when you felt the behaviour had control over your actions. What were the circumstances and how did it make you feel?

- How do you imagine your future self without the influence of the behaviour? What changes would you like to see in your overall wellbeing?

PART 2
BALANCING PURPOSE AND PLEASURE

In Part 1, we delved into the neuroscience of dopamine, looking at how it underpins many of our day-to-day, enjoyable activities. The focus was on the feeling of pleasure and anticipation. In Part 2, we explore the concept of *purpose*. What do we value? What gives meaning to our life? How do we find purpose? Are pleasure and purpose mutually exclusive, or can we strike a harmonious balance between them? (Hint: We can have both!) I'll guide you through exercises designed to help you identify your values, understand their origins and set goals aligned with them. But what does this have to do with dopamine? Well, in Part 3, I'll show you how to take a break from your dopamine-driven behaviours and replace them with values-based actions.

Chapter 8

What are values and why are they important?

Let me tell you about a 21-year-old client of mine named Ben. Ben told me he was experiencing symptoms of depression. He felt flat most days. He didn't get excited about much in life anymore. He slept more than he probably needed to and would sometimes take additional naps during the day. In his down time, he played video games.

In person, Ben was friendly and charismatic. You would have assumed he had a lot of friends, the kind of person who'd have a chat while waiting in line to get a morning coffee. His father was an accountant and ran his own successful business and Ben was halfway through an accounting degree. His family had always said that it would make sense for him to follow in his father's footsteps and one day take over the business. Ben was also a gifted athlete and had been involved in competitive long-distance running as a teenager, which he continued into adulthood. He'd played in a band with his best friend, but his friend moved overseas a year earlier and the band stopped.

On the surface, Ben seemed to be living a great life. And yet he wasn't happy. As I got to know him better, it became apparent that he wasn't living a life that was consistent with his values. He was doing a university degree which he didn't enjoy, and he had no real passion for a career in finance and accounting. He resented the amount of time and effort he put into his long-distance running but continued to do it because he had been good at it as a teenager. He'd won awards and it had become a part of his identity, at least in terms of how other people saw him. He'd been praised and recognised for his achievements. And he was chasing short-term gratification through playing video games, which he acknowledged was a way for him to 'tune out of life' and immerse himself in an alternative world.

I put the following scenario to Ben: imagine we could wave a magic wand that would make your life perfect. Exactly how *you* want it to be, where you didn't feel down or depressed. What would that look like? What would be different to now?

Ben struggled to answer these questions. He wasn't sure. He'd never really given it much thought. After all, he was only twenty-one. He did know, however, that there were parts of his life that he wanted to change. Over a period of time, we worked together to answer those questions, clarify his values, and work out how he could start to live life more in line with those values.

In this part of the book, I'll guide you through a few of the exercises Ben and I did together. I first helped him identify his values, then assessed whether he was content with how he was living life according to these values. Finally, we looked at what sort of changes he needed to make.

•

So, what exactly are values? Values are essentially beliefs that we have about what is important to us. Values give us a sense of meaning and purpose and serve as our guiding principles. They underpin our attitudes, our thoughts and our behaviours, and they are central to our sense of self and the lens through which we view and live life. I like to think of values a bit like a compass. Just like when we hold up a compass and it points us in the direction of our destination, values guide us in a particular direction throughout our lifetime.

Our values influence our daily choices, whether we are aware of it or not. They help us make major life decisions, such as which career path to follow, whether to stay or leave a relationship, whether or not to have kids, and where we choose to live. Values are what drive people to attend protest marches, go to church on a Sunday, donate money to charities and support friends going through tough times. Values help determine the trajectory of our future, they influence our relationships with others, and they play a crucial role in our wellbeing.

We all have values. Some of us might be very aware and conscious of them. Others might not. For some of us, we might only be aware of the obvious ones like family or health, but not the less obvious ones such as loyalty, grace or simplicity. Becoming aware of our values is a crucial first step to helping us make informed and personally meaningful decisions.

Values are different to goals. We never really 'complete' a value like we can complete or achieve a goal. For example, say that I hold the value of creativity (which I do). That value stays with me as I progress through life. Its priority might change as I move through life, however the value will be with me for a long time. Values are not goals, but they help us make meaningful goals for ourselves. More on that in Chapter 13.

Values are a core part of who we are and what we do. They are central to how we choose to live our life. Note the word 'choose'. This is important. Sometimes we automatically act in line with our values. At other times, we must consciously *choose* to do so.

Knowing what our values are can help us in so many ways. It can help us communicate with others. Values allow us to express thoughts and ideas from a place of clarity and understanding. They help us resolve conflicts and make hard choices when we're faced with difficult decisions. They help us plan for both the long term and the short term. They help us form friendships and relationships and are crucial in building strong interpersonal foundations. They provide us with a guide about when to compromise and when to stand firm. And they act as a filter, helping us focus and prioritise what is most important.

Clarifying our values is an incredibly useful thing to do. Take, for example, making decisions about health care. Most people find this challenging. Doctors are trained to offer their clinical expertise, while also honouring the requests and wishes of patients, but people don't always know what's important to them, which makes the decision-making process hard. Clarifying our values has been shown to help people when faced with hard decisions.[1] It gives them some autonomy in the decision-making process. It helps parents make gut-wrenching decisions about palliative or end-of-life care for their dying children.[2] It helps them make decisions about undergoing genomic sequencing[3] for their newborn child, or when a foetus[4] is diagnosed with a life-threatening condition. It's during the toughest of times that knowing what our values are is so crucial.

Most people work out a relatively stable system of values during childhood and adolescence.[5] As we get older, our understanding of moral reasoning – what is 'right' and 'wrong' – becomes more sophisticated. While values are relatively stable and don't easily change, as we move through life, certain experiences call upon us to reflect and review our values. We may acquire new ones, or shed others. At other times our values may change in terms of order of priority. Some might have the same importance throughout our life. Life transitions, the influence of other people, and ageing all affect our values.[6]

Young adulthood is a dynamic stage of life. During this time, we encounter many major life transitions and new social roles, such as moving out of home for the first time, starting (or ending) a first serious relationship, making decisions about our career, beginning tertiary education, entering the workforce and making new friends.

For me, the things I valued in my twenties are not necessarily the things I value now. In my twenties, friendship and leisure were more important than my career. Now, in my thirties, it's the other way around. In my early twenties, I didn't realise that I valued adventure. But road trips through remote parts of outback Australia, a helicopter ride through the Valley of Fire in the US, horseback riding on ranches and sand-boarding down desert dunes proved how much I value the thrill of adventure and new experiences. But until I had done all those things, I didn't truly know how important they were to me.

The roller-coaster of life comes with constant change and is filled with both joy and tragedy. Some life changes are planned, while others are thrust upon us. Relationships begin and relationships end. Children are born. People move houses, suburbs, cities and countries. Careers progress and careers change. We get new jobs and we lose others. Some friendships blossom while others end. People become unwell. Some recover and some don't. We celebrate new life and we grieve lives lost.

As life progresses, we'll all be confronted with illness and death. These will be times of significant emotional turmoil, but they can also serve as points where our values are tested and reflected back to us. We often hear people who have survived serious illness say that the experience changed their outlook, particularly if they were faced with the possibility of dying.

Good and bad experiences reflect our values and priorities, what is important to who we are and how we choose to live. Furthermore, living life in alignment with our values has positive effects on our general wellbeing. People who report that they're fulfilling their values also report a higher level of wellbeing.[7] Even the simple act of thinking or writing about fulfilling our values can contribute to a greater sense of wellbeing.[8]

In fact, values are so important that the German curator and cultural translator, Jan Stassen, created a 'museum' dedicated to them! People are asked to send in an object to the museum which is attached to a story that represents an important value to them. The objects are said to be 'witnesses and relics' that remind us of what is important. Part of this process is about bringing our values to the forefront of our minds, as they can very quickly and easily get lost in the background. Jan Stassen says: 'Values are the social glue that hold us together, with all our beautiful differences.'[9]

So, what do values have to do with dopamine? Well, nothing really! And that's the point. Actions based on values and those based on short-term dopamine hits are often at odds. That makes values important to consider and discuss, as the craving for dopamine-driven behaviours can divert us from engaging in activities that are meaningful and aligned with our values. When I think about how I'd spend my evening in a way that is aligned with my values and what is most meaningful, I'm not going to actively *choose* to spend it scrolling through TikTok. The dopamine hook of reinforcement means that five minutes of scrolling can easily turn into fifty minutes and next thing I know, I've spent much longer online than I anticipated! Similarly, my client Ben who was feeling depressed and dissatisfied chose to spend his spare time playing video games, rather than engaging with more meaningful pursuits. Games are fun every so often, but they don't help us live a fulfilling life in the long run. These are examples of how the pursuit of uninhibited pleasure pulls us away from our core values, and from doing more meaningful things in life.

In the next chapter, we'll do a challenge that is both fun *and* enlightening to identify our core values. This will help us when we take a break from our chosen dopamine-driven behaviour or substance. It'll allow us to re-evaluate our priorities in life and help us make choices that are aligned with what is fulfilling and meaningful.

LET'S REFLECT

- What are some aspects of your life that you find personally meaningful and fulfilling?

- Think about a recent decision you made. Which of your values may have helped you make that decision?

- How might your values have changed throughout your life?

- Can you identify any significant life events that prompted a shift in your values?

Chapter 9

Identifying your values

Now for the fun part – or at least I think so! Identifying our values.

We are going to split this into a few different activities. First of all, I'd like you to brainstorm a list of values. Consider the things that are valuable and important to you. It doesn't matter if you're not living out these values at present. For example, I might value 'adventure', but do nothing in life that reflects that value. That's okay! Still write it down. We'll take action later.

After brainstorming, I'll give you a list of values to help you identify additional ones that you may not have thought of. (Note: People always identify more values once they look at the list – it's long!) But it's important to brainstorm first. So take a pen and paper, or make a note in your phone, and complete Step 1 before having a peek at the list over the page.

After that, we'll do an imaginary exercise. I'll get you to close your eyes and reflect on certain parts of your life. This will help you see if your reflections are in line with the values you have identified.

Finally, we'll narrow down your values to somewhere between five and ten and use these to create some goals and an action plan.

Ready? Let's go!

Step 1: Brainstorm your values

Below are some questions designed to help you identify your values. You don't have to answer each and every question. Take a moment to reflect and write down the words that come to mind. Try to think of single words that describe broad categories (for example, 'religion'), rather than specific examples or actions (for example, 'going to church every Sunday').

What things are important to you?

What things are meaningful to you?

What are you passionate about?

What topics get your heart racing and make you speak a little louder when they come up in conversation?

Recall times when you have been so absorbed in what you were doing that you hardly noticed time passing by. What were you doing and what values might that relate to?

What qualities do you value in those around you?

What qualities do you strive to hold yourself?

When you look back at your life, what are the moments you are proudest of? What values are relevant to those moments?

When do you feel most content?

Hopefully you've been able to come up with a list of at least five values. To ensure that you have identified values and not goals, remember that values *don't have an end point*. For example, *health* is a value, but *going to the gym five times a week* is a goal. We can never 'finish' or 'complete' a value like health. It'll be with us for our entire life.

If there are some goals that are particularly important to you, consider the exercise in reverse. What values might be driving that goal? For example, if you have the goal to pay off the mortgage on your house, what value (or values) does that goal tap into? It might be that you value financial security. You might value wealth. You might value hard work and success. If you go to the gym five times a week, you might value health, but you might also value your physical appearance. There can be multiple values that drive certain goals. If you are more in touch with your goals rather than your values, try this reverse exercise to identify what the core drivers are. This will help you to set even more goals that are aligned with your values. It may also help you realise that you have some goals which are not in fact grounded in your values.

Step 2: Choose from the values list

Over the page, you'll find a long list of values. The trouble most people have is not that they can't identify enough values, but that there are *so* many that feel important! Try to be discerning. Don't choose every value on the list, even if they all feel important. That would make this exercise almost meaningless! Just choose the most important ones.

Although the list is lengthy, it's by no means exhaustive. If you've identified a value in Step 1 that is not on this list, don't worry. Hold onto it if it feels appropriate. Some of the values may also be similar – for example, 'fortitude' and 'courage'. Go with the value that resonates most for you, but don't double up if the similar but alternate description doesn't add anything different.

If you've completed that task, congratulations! It's not easy and often brings up *a lot* of self-reflection. Hopefully you now have a list of quite a few important values. Remember, these don't necessarily need to be things that you are acting on in your life, but rather things that, in theory, are important to you.

Acceptance
Accountability
Accuracy
Achievement
Adaptability
Adventure
Altruism
Ambition
Appreciation
of beauty and
excellence
Assertiveness
Attractiveness
Authenticity
Authority
Autonomy
Balance
Beauty
Belonging
Boldness
Bravery
Calm
Career
Caring
Certainty
Challenge
Change
Charity
Citizenship
Clarity
Cleanliness
Collaboration
Comfort
Commitment
Community
Compassion
Competence
Concentration
Confidence
Connection
Contentment
Contribution
Conviction
Courage
Courtesy
Creativity
Curiosity

Dependability
Determination
Dignity
Discovery
Diversity
Duty
Ecology
Efficiency
Endurance
Environment
Equality
Excellence
Excitement
Fairness
Faithfulness
Fame
Family
Fidelity
Fitness
Flexibility
Forgiveness
Fortitude
Freedom
Friendship
Fun
Generosity
Genuineness
God's will
Grace
Gratitude
Growth
Happiness
Health
Helpfulness
Honesty
Honour
Hope
Humility
Humour
Imagination
Improvement
Inclusion
Independence
Initiative
Inner peace
Insightfulness
Integrity

Intimacy
Joy
Judgement
Justice
Kindness
Knowledge
Leadership
Learning
Legacy
Leisure
Logic
Loving
Loyalty
Mastery
Mindfulness
Moderation
Monogamy
Nature
Non-conformity
Nurturance
Openness
Optimism
Order
Passion
Patience
Patriotism
Perseverance
Perspective
Pleasure
Popularity
Posterity
Power
Pride
Prudence
Quality
Rationality
Realism
Recognition
Reliability
Religion
Reputation
Resourcefulness
Respect
Responsibility
Restraint
Risk-taking
Romance

Safety
Security
Self-acceptance
Self-control
Self-discipline
Self-expression
Self-knowledge
Selflessness
Sensitivity
Serenity
Service
Sexuality
Sharing
Simplicity
Sincerity
Solitude
Spirituality
Spontaneity
Sportsmanship
Stability
Status
Stewardship
Success
Sustainability
Temperance
Timeliness
Tolerance
Toughness
Tradition
Tranquillity
Travel
Trustworthiness
Truth
Understanding
Uniqueness
Usefulness
Valour
Victory
Virtue
Vision
Vitality
Vulnerability
Wealth
Wellbeing
Wisdom
World peace
Zest

Step 3: Imagine 'future you'

This is where we get creative and use our imagination.

I love the concept of 'future me'. Whenever I'm facing a problem that I can't seem to solve, I tell myself, 'That's future Anastasia's problem'. Sometimes, future me looks back at past me and curses my disorganisation. Other times, future me looks back with thanks and gratitude. The notion that there's a version of myself that doesn't exist yet – a version that's a little bit older, a little bit wiser and hopefully a little more capable than I am right now – brings me comfort. It allows me to defer problems and sit with the thought that I don't need to solve everything right here, right now. The notion of future me also inadvertently reminds me of change. More specifically, that *we* can change. That I have a life ahead of me and I have choice over how I live it.

So let's turn on that imaginative and creative part of the brain and take a moment to imagine our future self. (This isn't a new exercise I've made up. Lots of therapists before me have used this as a means of helping clients to clarify their values and priorities.) Imagine yourself as an eighty year old. Now imagine a setting where you feel calm and content. A place of serenity. Perhaps you're by a peaceful lake, or at the beach listening to the sound of the waves. Perhaps it's in the comfort of your bedroom, or a favourite place you like to go that has fond memories.

Now, take a moment to think back on your life and ask yourself these questions:

What would you like to be able to think about yourself?

How would you like to feel about yourself and the person you have been throughout your life?

How would you like to have treated and interacted with others?

What would you like to be able to think and say about the life that you lived?

Are there things you would have liked to have achieved? What are they?

How would you like other people to describe who you have been and how you've lived?

Spend some time thinking about these questions. They'll provide insight into those longer-term values that are important to us. Once you've given it some thought, write down some notes. These will help you to understand your values and what's important.

Perhaps you wanted to be able to look back on your life and think, 'I was a kind and generous person'. Or perhaps, 'I was successful in my career'. You might aspire to be someone who fought tirelessly for equality, justice and human rights, trying to effect positive social change. Or maybe you want to feel pleased that you lived an unconventional life, full of adventure and spontaneity. You might want to feel content with your devotion to your religion or spirituality. You might realise that leaving a legacy and something that transcends and exists beyond you and your finite life is important. This could be a body of work, some art, or even children. There's no right or wrong in this exercise, only information about yourself and what is fundamentally meaningful and valuable to you. Use this either to seek out additional values or to confirm the values you already thought were important.

As a side note, thinking about ourselves as an eighty year old can be confronting. People who have anxiety or avoidant tendencies may be reluctant to do visualisations like this. Notice that discomfort. It might raise questions for us like 'What if I don't become the eighty year old that I hope to be? What if I don't live the life I hope for? What if I don't always act kindly and generously, I don't progress in my career, or I never get to leave behind an enduring legacy?'

To that I say this: no-one gets it 100 per cent 'right' and no-one lives a perfect life. No person, regardless of experience, age or wisdom, ever achieves perfection in navigating life's challenges. There are always things we will look back on and think that we could have done better, more effectively, more kindly and compassionately, or more successfully. That's okay. Hindsight is, as they say, twenty-twenty vision. The realisation that a perfect life is unattainable is a liberating concept. It allows us to cultivate compassion and under-standing for both ourselves and others.

Don't let a 'fear of failure' be a barrier to this type of reflection. Thinking about our future selves and the ideal life to have lived gives us the best chance of fulfilling that ideal. We won't get it 100 per cent right, but we have a better chance of living a life of contentment and fulfilment if we confront and embrace these reflective questions, rather than avoid them. *Nothing changes if nothing changes.* This is our moment to reflect deeply and think about how we can make realistic changes to give ourselves the best chance of living the life we truly hope for.

Step 4: Narrowing our values

Okay, so, Steps 2 and 3 were the easy parts! Unfortunately, this is the harder part (or at least that's what my clients tell me). In Step 4, we are going to *narrow the focus.* Your task here is to try to reduce your list of values to a maximum of ten (fewer than ten is fine too). These will become your fundamental, core values. The ones that are most important to you. They are non-negotiable.

Identifying *too* many values isn't particularly useful because there is only so much we can do and so much change we can make. I once gave a client the values list to take home, reflect on, and then select the most relevant. In the next session they said they had managed to cross about ten off the list but all the others felt important! We had a laugh about how it would be impossible to use the ninety or so values that remained productively to make meaningful changes. We then started the process of narrowing them down.

If we have a list of twenty or more values and try to consistently live life in line with them every single day, we're setting ourselves up for disappointment. It's not to say that we discard the others, but rather that it's far more useful and productive to focus on a select number to start with. They'll soon become second nature and you'll have more mental space to then focus on some of your other values.

I hope this chapter has enabled you to reflect on and identify your core values. They'll shape your identity. We've also discussed goals and aspirations, and thought about our future selves to understand what we want our lives to ultimately be about. Keep those core values close to you as you read on. They'll help you to choose intentional values-driven actions, rather than unintentional dopamine-driven behaviours.

LET'S REFLECT

- Are there any values you identified that surprised you?

- Think about activities that promote immediate gratification. Do they align with your core values? Can you identify any times where the pursuit of short-term pleasure interfered with more meaningful actions?

- Can you recall some specific moments where you felt completely absorbed and fulfilled? How do the values you identified align with those moments?

- Some values feel more important than others. Which on your list are non-negotiable and fundamental?

Chapter 10

Where do my values come from?

Now that you have a list of your core values, take a moment to consider the following questions:

Where do my values come from?

Are they really my own?

Or are the values I have selected ones which my family, friends or society tell me are important?

A lot of the time, we internalise the values of others. Now, that's not a bad thing! It can be useful to look to friends, family or well-known people to get a sense of what they hold as important and what gives them a sense of meaning and purpose. But we need to think about where our values come from to ensure that they really are true reflections of what is important to us, and not just what society tells us is important.

Take 'hustle culture', for example. In recent times, we've seen a rise in the value placed on extreme hard work and success. Hustle culture is all about the 'rise and grind', pushing yourself to the limit to achieve as much as you possibly can. Hustle culture has little regard for rest, self-care or work–life balance. Instead, the focus is squarely on productivity, ambition and success. Society tells us this approach is valuable and important and those who adopt this set of values will be rewarded for their achievements. However, it can also go terribly wrong. Hustle culture is a recipe for burnout.

My client Belinda held a senior role as an investment banker. She spent the early years of her career working exceptionally long hours. She'd be at the office late into the night, often not leaving until midnight, then waking up the next day and doing it all over again. Her weekends were often spent finishing tasks she didn't get to during the week, or purely trying to rest and recover. She was having trouble falling asleep and staying asleep, and she couldn't fall pregnant. Belinda lacked work–life balance.

When I saw her, she'd taken a week's sick leave to deal with serious depression, burnout and the effects of chronic, high-level stress. That one week turned into six months of leave, and eventually a new job. Belinda's case was a good example of hustle culture gone off the rails. Society told her that working hard was important in order to get ahead. Her employers and mentors told her that sacrificing her weekends and sleep were important if she wanted to climb the corporate ladder. Her family told her that a career in finance was important, as her father had held a powerful position as a partner. Belinda received the messages, both directly and indirectly, that her work and career were the key things in life. Success in her field was highly valued and success only came about through hard work. But not only was the 'rise and grind' a disaster in the making for Belinda, she soon realised that it *wasn't* part of her core values at all. She'd internalised the values of others: society, her workplace, corporate culture and even her family. And she'd forgotten her own.

The greatest joy for me came in our penultimate session, when Belinda told me she was pregnant! She and her husband had been trying for over a year without success. It was no great surprise that when she finally stepped away from her stress-fuelled job it happened for her.

I often say to clients at our last session (with the utmost respect and care) that I hope *never* to see them again, because that means they are progressing well in life. I'm quick to add that my door is always open, should the need arise. I'm happy to say I haven't seen Belinda for a few years now, so I hope that means that motherhood is treating her well and she's put the rise and grind of hustle culture behind her for good.

Our cultural upbringing, as well as the cultural groups with whom we identify, also influence our values. Some cultures value 'individualism' while others value 'collectivism'. In individualistic societies (such as in the West) autonomy, independence and one's personal pursuits are seen as important. Personal interests and goals are influential in directing behaviour. On the other hand in collectivist societies (such as in Asia) cohesion, social interdependence and harmony are more important. The needs and goals of the collective take priority over those of the individual. In collectivist cultures, it's common for multiple generations of family members to live together or in close proximity. Adult children often take care of ageing parents at home. In individualistic societies, while family support is important, independence in life and separate living arrangements are more common and celebrated to a greater degree.

Religion or spirituality are also a source of values for many people. Religion can provide a framework that dictates how to live a meaningful and purposeful life, and different religions offer different sets of values, principles and moral guidelines. They give people ethical guidance, a context for understanding meaning and purpose, instruction as to how to treat others, and a compass to navigate justice, fairness and personal integrity. Spirituality, which is not associated with any particular religion, allows for the exploration of the self, as well as offering a higher purpose. While individuals can define their own spiritual path, the concept requires self-reflection and consideration of something greater than ourselves.

Cultural groups aside from race and ethnicity also influence our values. Indeed, the group's foundational values might be the reason we are drawn to them in the first place. For example, people who identify with an artistic subculture may value freedom of self-expression. Hip-hop culture, for example, values authenticity and social justice activism, while the emo subculture values emotional expression. People living in rural areas might value a simpler way of life and farmers might value sustainability of the land. The values of those who serve in the defence forces might be centred around duty, loyalty and mateship.

With our ability to move around the world, it's not uncommon for people to experience a 'culture clash' and a subsequent clash of values. Migrants, the children of migrants, refugees, expats and international students often find it hard adjusting to the culture and values of their new homes. For children raised in the West by parents who were brought up elsewhere, there's often tension between the values held by the family and the values held by society. Negotiating these clashes can be a delicate process and can lead to people questioning their values and identity.

These clashes of values also occur on a smaller scale within families. Think back to Ben in Chapter 8. He was feeling dissatisfied partly because his family valued sporting achievements, discipline and following in his accountant father's footsteps above all else. As we worked together, it became clear that the values Ben thought he had weren't in fact his: they were his family's. He soon discovered that he had some different values. He valued freedom, spontaneity and fun more than self-discipline and sporting success. Living his life in a way that was consistent with his family's values and not his own made him depressed and dissatisfied. Unearthing his own true values allowed us to address the root cause of his struggles. Once that was done, we worked towards aligning his life choices.

The origin of our values and how 'authentic' to us they are form part of our personal identity. They then influence the development of our *social* identity (that is, the portion of our self that is derived from how we perceive ourselves to fit within relevant social groups).[1] The American child psychoanalyst, Erik Erikson, was known for his work on identity formation and the psychosocial development of human beings.[2] He wrote that most people experience 'identity crises' and conflicts throughout their lives. During these crises, they try to figure out who they are, often based on social and employment identities. The crisis is resolved when they consider their various goals and values, and accept some and reject others. In this way, they begin to understand who they really are.

Think of your identity like a pie chart. The different segments of the pie represent characteristics that are important to your sense of self. These are usually values driven. Now, not all these segments may be values, but if they are important enough to make up your identity (and be included in your pie chart) then they will likely reflect your values.

For instance, I'm of Greek background and part of my identity pie chart would be having Greek heritage, which links into an important value of mine – culture. Other pieces in my pie chart would be seeing myself as a daughter (value = family), a friend (value = connection), a psychologist (value = helping others), a musician (value = musical creativity) and a traveller (value = new experiences/adventure). The components of my identity speak to the values I hold.

In this chapter I've provided a lot of examples of values that are influenced by other people, groups, cultures or society. It is important to ensure your values are truly your own, and not just those you've adopted because of the influence of others. Consider which are personally meaningful. You may find that you have inherited them from your family or culture. You might really admire those values. If that's the case, fantastic! All we need to do is make you more aware of the wider context so that you can ensure you're choosing what's right for you.

LET'S REFLECT

- Where do your core values come from?

- What influence have your family, your upbringing, your culture, and your local community or society had over your values?

- Do you share similar values to those around you? Do your shared values contribute to the strengthening of your relationships?

- Have you ever experienced conflict or difficulties in a relationship due to a difference in values?

- What are the segments in your identity pie chart? How would you describe yourself to someone who has never met you? What values do these parts of your identity reflect?

Chapter 11

Assessing our alignment

By now I hope you have a list of core values that are truly your own. But there's more work to do! Merely identifying those values is unfortunately not enough to achieve fulfilment in life. *Knowing is one thing, doing is another.* The next step requires some deep introspection to assess whether the life you are living is truly aligned with your core values.

As most of us struggle to find the delicate daily balance between pleasure and purpose, it's okay to have room to improve. Realistically, we never get it right 100 per cent of the time. As clichéd as it sounds, life really is a journey (and one that requires constant reflection and constant adjustment). It's never set and forget! Reflection is important for building self-awareness of our thoughts, emotions, beliefs, behaviours and choices, so that we can be conscious and deliberate in our decisions and actions. Reflection helps us clarify our goals and priorities, strengthen our sense of identity and build meaningful relationships. And it allows us to make choices that carefully balance our own needs with the needs of others.

According to researchers Monika Ardelt and Sabine Grunwald, self-reflection allows for increased awareness and openness to ourselves, others and the world. It fosters human development, personal transformation and, ultimately, a better society.[1] Therefore, our capacity to reflect affects not only us, but the world! We all have room for personal growth and structured reflection allows us to identify where we can make truly meaningful changes.

We can test if our values and actions are aligned by analysing our daily behaviours. Often people think that making meaningful change will require some sort of big action or grand gesture. But in fact, the most sustainable and successful change usually comes from *small* but *consistent* shifts in behaviour. It's inevitable that people unintentionally drift away from their values. This can happen for all sorts of reasons, including societal expectations, external pressures, or simply the hustle and bustle of daily life. Values are not always at the forefront of our mind. Pausing to assess the alignment between our values and our actions is the vital next step.

To help you determine whether you're driven more by pleasure, or more by purpose, let's run through some questions. For each value, I want you to first ask yourself:

Am I, for the most part, living my life consistently with this value?

The answer might not be as simple as 'yes' or 'no'. It might be, 'Yes, when I'm with certain people', and 'No, when I'm with other people'. Or, 'Yes, but only when I remember.' Or, 'Not really because I'm busy.'

The second part of the reflection involves you asking yourself this crucial question:

What things might get in the way of living life consistently with this value?

Again, this is not a straight-forward question with a straight-forward answer. There might be many things that get in the way of acting in alignment with your values. For example, you might not have the time, the energy, the money or the mental capacity. You might forget about a particular value or the actions associated with it, especially if it's not at the forefront of your mind. For example, the value of family might be easy to remember and action if you're surrounded by family constantly whereas the value of outdoor adventure may be easier to forget if you're stuck in an office five days a week!

The barriers between you and your values may go even deeper. Perhaps you're avoiding certain values because they remind you of painful or difficult times in your life. I remember working with an older lady who had attended church her entire life but stopped going because of the painful reminders of her husband's funeral. She experienced inner conflict because church and religion were still very important to her, but the memories were hard and brought up a lot of emotions. We see similar patterns of avoidance among people who value relationships but have experienced significant hurt and heartache. Having a romantic relationship is important to them, but they avoid dating or forming deep connections for fear of getting hurt again. A person might be dissatisfied in their chosen career, but they avoid changing direction because it feels too hard and overwhelming.

Your own thoughts and beliefs can also stop you acting in accordance with your values. For example, you might value career success, but believe that you're 'not good enough' to succeed, or destined to fail. This belief might prevent you from pursuing goals aligned with your value. You might not apply for the job or promotion out of fear that you won't get it. Or you might value health but tell yourself, 'I don't have the discipline to exercise regularly', or 'I have a genetic family history of health problems, so what's the point?'. Or you might value community, but see yourself as an 'outsider' who doesn't fit in, which blocks you from engaging and forming connections, only further consolidating the belief that you don't fit in. It can become a vicious cycle.

Another common barrier is that most of us have *a lot* of things in life that are important to us. It can be challenging to find the time, effort, energy and resources to do them all consistently. In fact, sometimes engaging with one particular value might mean giving up time doing something else. If I value charity and want to spend time each month volunteering at the local soup kitchen, I may need to give up some time spent with friends or family, or time I might have spent exercising or working. We have a limited amount of hours in the day and a limited amount of energy too! Sometimes engaging with one value means we sacrifice another. This is why making the transition from identifying values to actual behavioural change is so challenging. It's a constant balancing act.

While we're focusing on meaningful values, we also want to make sure we have moments of fun. This is why I encourage you to select only five to ten values to focus on (at least to start off with). If you chose twenty from the list in the previous chapter, you'll find it very hard to meaningfully implement changes and you might get disheartened, feel overwhelmed and then give up! Remember, *small consistent change is good change.*

Remember Ben from Chapter 8, the young man who was depressed and not living life according to his values? He identified his top five values, in no particular order, as spontaneity and fun, family, adventure, creativity, and kindness. We then applied a two-question approach to examine whether his daily life reflected these values.

For the value of family:

Am I, for the most part, living my life consistently with this value?

Ben agreed that for the most part he was living in alignment with prioritising family.

What things might get in the way of me living life consistently with this value?

Ben identified that feeling pressure from his family sometimes affected his desire to spend time with them. And being busy with assessments at university affected his ability to have the mental or emotional capacity to give back as much to his family as he would like to.

We then reflected on the value of creativity:

Am I, for the most part, living my life consistently with this value?

Ben's response was that he wasn't doing anything creative. He had previously played in a band with his best friend, but this stopped when his friend moved overseas a year earlier.

What things might get in the way of me living life consistently with this value?

Ben found that playing music sometimes produced uncomfortable emotions. While he loved music, it connected him deeply to his feelings. He spent time trying to avoid these feelings and 'tune out' by playing video games. He was sad that his best friend had moved away, and he was feeling more disconnected ever since the move. In essence, the uncomfortable emotions were serving as a barrier to Ben playing music or engaging with his value of creativity.

Musical creativity is a value that I struggle with in my life, but for different reasons to Ben. I started learning the piano at age three (my mother figured I didn't have anything better to do with my time, so why not start piano lessons!). While at school, I also learnt the violin and viola, and took lessons in singing, guitar and flute. I even learnt the bagpipes at one point thanks to my school's Scottish heritage. I was in lots of choirs and orchestras and took every opportunity to perform. Being on stage felt exhilarating. As I got older and pursued my studies and career in clinical psychology, it became harder for me to play as much music as I would have liked. Now, the barrier for me is that my life is not only busy, but structured. Much of my research work is analytical and scientific, and I've found that I need to schedule and block out time for creative musical pursuits.

My solution has been to keep in touch with friends who are musicians and find opportunities to play with them whenever possible. I've moved from just playing classical music to a wider range of genres. Many of my friends are country musicians, which means that through them, I get to play some boot-scootin', honky-tonk tunes at gigs every so often. It's a lot of fun! I go to concerts and shows, which also help me to feel connected to my value of musical creativity. And I've taken a special interest in working with people in the music industry in my day job, which has allowed me to provide mental health support at music festivals and events around Australia. We can only effectively overcome barriers if we know what those barriers are. I realised that my value of musical creativity didn't naturally 'fit' into my life anymore, so I had to find ways to make it fit.

If your answer to the first question of 'Am I, for the most part, living my life consistently with this value?' is 'yes', then great! Keep doing what you're doing because it's obviously working. But if your answer is 'no', or not as much as you'd like to be, then we have some work to do.

LET'S REFLECT

- What are the challenges or obstacles that get in the way of you living a life in alignment with your values?

- Are the barriers you identified internal or external? Are there beliefs or difficult emotions preventing you from fully engaging with your values?

- How does it feel when you experience a disconnect between your actions and your values?

- Are there times you can recall where self-reflection has contributed to you building your own self-awareness? Has self-reflection helped you to clarify your goals?

Chapter 12

Navigating the fork in the road

At every moment in time, we have choices to make. In fact, life is a series of choices, both big and small. We can't control everything, but there is certainly a lot that we *can* control. Each day, we choose whether to hit the snooze button when the alarm goes off and have a sleep in, or get out of bed early. We choose what to eat for breakfast, what to wear, and whether to take the stairs or the elevator up to the office. Some of these choices are minor and have little impact on our life. But some are major and profoundly influence the path we travel. Indeed, some can truly be life-changing. Whether to quit our job, try for another round of fertility treatment, get married, sign divorce papers or undergo a major medical procedure all affect our journey in dramatic ways.

Many great philosophers have written about the concept of choice. Aristotle believed that, 'It is our choice of good or evil that determines our character, not our opinion about good or evil'. He also wrote that it was 'choice not chance' that determined our destiny (I only partially agree with that). The Swiss psychiatrist, Carl Jung, put it this way, 'I am not what happened to me, I am what I choose to become'. The 'free will' philosophers argue that humans have complete autonomy to make their own decisions, while the 'determinists' state that human behaviour is merely a result of internal and external forces acting upon us. Even J. K. Rowling in the *Harry Potter* series had something to say on the matter. According to Dumbledore, 'It is our choices, Harry, that show what we truly are, far more than our abilities.'

And let's not forget the central paradox of choice – the notion that having too many options can make us stressed, dissatisfied or vulnerable to a state of 'decision paralysis' where we end up choosing nothing at all! Ever been to a restaurant that had too many dishes on the menu? It makes it much harder to decide what we want to eat. In another contemporary example, too much choice is a big issue for those who use dating apps. Users have access to a vast pool of potential matches. While more options might initially seem wonderful, too much choice soon leads to less satisfaction, decision paralysis, and constantly searching for something – or someone – better.

On the other hand, choices give us power. And we *always* have a choice, even when we feel like we don't. If we think we don't, or don't feel confident that we can make a choice, we get stuck. The reality is there are certain things in life that are out of our control. There are aspects over which we have no agency, and all we can do is accept them for what they are. The weather is a good example. Try as I might, I cannot change the weather forecast for tomorrow to be sunshine instead of rain. Death is another aspect of life over which I have no control and am forced to accept. One day I will die and that is a reality I cannot escape. I can't change the past, nor can I completely change the future. I can't change other people, and I can't change their actions (despite sometimes really wishing I could!). These are all things over which I have no control.

What I can control is what *I do*. I can choose my thoughts, my beliefs and my actions. I can't change the weather, but I can choose to pack an umbrella if the forecast is for rain. I can't escape death, but I can choose to live life to its fullest every day. I can't change the past, but I can choose to change my relationship with it. A lot of the future is out of my control, but I can make choices every single day that influence the direction my life takes. And while I can't change other people, I can choose how I respond to them.

In philosophical terms, what I'm describing is what's known as the 'dialectic' of acceptance and change. Dialectics describe two things that seem opposite, but are actually both true. The idea has its origins in ancient Greek philosophy. It was a form of reasoning or debate where truth was discovered by considering opposing or conflicting points of view. To give you an example, I can be in a room full of people and yet feel completely alone. The two scenarios might seem like complete opposites, but both are in fact true at the same time. Another example might be a 'bittersweet' experience that is both happy and sad. The same principle applies to acceptance and change. In any given situation (and I am yet to find one where this doesn't apply), there are certain elements that are within our control that we can change. At the same time, there are certain elements that are out of our control that we cannot change and must accept.

The fundamentals of the dialectic of acceptance and change are described in the famous *Serenity Prayer*. The *Serenity Prayer* was thought to have been written by the American theologian, Reinhold Niebuhr, in the 1930s as the ending to a longer prayer. Since then, it has been popularised by Alcoholics Anonymous and other 'twelve-step' recovery programs. The prayer says, 'Grant me the serenity to accept the things I cannot change, the courage to change the things I can, and the wisdom to know the difference.' This is at the heart of the dialectic of acceptance and change: we need to accept the things we cannot control and work towards making the changes in our life where we do have control, while understanding that these two things are different.

Do not make the mistake of thinking 'acceptance' is the same as 'approval'. People sometimes get stuck on the notion that to accept something means that they approve of it. Acceptance is *not* approval. Rather, to use a modern phrase, it is acknowledging that 'it is what it is'; I do not necessarily like it, approve of it, or agree with it, but I cannot change it. For example, I didn't like the fact that there was traffic this morning on my way to work. In fact, I found it frustrating and inconvenient, but I must accept it for what it is, because I cannot change the fact that the roads are crowded. I don't necessarily approve of the outcome of an election, but as I cannot change it I am forced to accept it for what it is. Acceptance is merely me acknowledging that something is out of my control.

So why the focus on choice and control in this chapter? Well, every day, we have the opportunity to make a choice. In psychological terms, we call this a 'choice point' and it's a component of a treatment called 'acceptance and commitment therapy'.[1] The choice point essentially represents a moment in time when you can choose a behaviour that moves you *towards* your values, or a behaviour that moves you *away* from them. Naturally we want to do the former! Why? Well, choosing an action that moves you *towards* your values takes you in the direction of being the person you want to be and living the fulfilling life you aspire to live. I like to imagine the choice point a bit like a fork in the road. You have the option of one of two paths. The choice point analogy allows us to pause and consider our values in a given situation before responding to it. The pause creates space for mindful self-awareness and allows people to choose behaviours that align with their deeper values and long-term goals.

Now, this may seem like a no-brainer, but it's actually very easy for us to be diverted from our values! A move *towards* them takes conscious and deliberate effort. A move away might be due to automatic habits that we quickly and easily adopt. Consider what might pull you away from your values. (Hint: It'll probably be one of those dopamine-related activities or behaviours we discussed earlier!) Life is full of difficult situations and we all experience unhelpful thoughts and unpleasant emotions. Our own internal experiences, as well as external influences, can make it hard consistently to select the *towards* path. And there'll be times when we just don't *want* to follow the towards path. For example, I know eating those delicious, sugar-coated doughnuts from the café at my clinic aren't great for me. They don't really align with my core value of health. And yet I still eat them! I'm choosing a move away from my value for the purpose of pleasure – and that's okay every once in a while. It's all about balance and moderation. It only becomes a problem when I cease being aware of my conscious choices and *automatically* take the 'doughnut path' and do so in excess.

So remember, every moment is a choice point. You chose to open this book and you're choosing to keep reading (I hope). Think about what values underpin these actions. Maybe you value self-reflection and your mental health. Maybe you value finishing what you started and not stopping reading a book halfway through. Or maybe you value knowledge.

As I say, we are constantly making choices – some trivial and others significant – but they all add up to create the life we'll look back on when we're eighty. We want to ensure that we are consciously making decisions that will result in satisfying and rewarding reflections when that time comes. Choices are good: they give us agency and help us maintain a sense of control over the things that we can change.

In the next chapter, we'll take a closer look at some 'towards' moves, those actions and behaviours that take us on the path to fulfilment. Knowing our values is one thing, but we also need to understand the goals and behaviours that underpin them.

LET'S REFLECT

- Reflect on a recent situation in your life where you were faced with a 'choice point'. How did you decide which path to take?

- Think about a situation where you had to accept something that was out of your control and that you couldn't change. How did you navigate the process?

- Recall a time you came to a fork in the road and automatically walked down the familiar, habitual path. What would have helped you stop and consider the options? How might the outcome have differed if you'd paused to reflect on your values?

- Consider the 'dialectic' of acceptance and change. Has there been a situation where you have had to simultaneously accept certain aspects while also making additional changes?

Chapter 13

Setting goals in line with values

In this chapter, we start to focus on *change*. By now, we've identified and clarified our values, narrowed them down to a select few, and reflected on whether we're actively living our life in line with these values. Now we'll take this one step further and move from the 'conceptual' to the 'practical'.

Unfortunately, just knowing what our values are is not enough. A person may have crystal-clear clarity around their values and priorities and yet live a life that is completely inconsistent with this knowledge. If our actions do not align with our values, we'll feel dissatisfied. We may even be susceptible to more serious mental health concerns. We'll have trouble making major decisions and we'll spend a lot of time stressing or ruminating, never feeling confident with the path we're choosing.

Simply knowing that I have the value of adventure is important, but I need to put it into practice. What exactly does adventure mean to me anyway? Does it mean I sell my furniture, pack up my belongings and go backpacking through Europe? No, but it might for someone else! For me, adventure means discovering new places, exploring the countryside, road-tripping through the outback, and embracing different cultures, landscapes and cuisines. For others, it might mean bungee jumping, sky diving, climbing Mount Everest or buying a ticket on a Virgin Galactic ship and heading into space!

Unless I define what 'adventure' means in practice, I cannot live a life in line with it. Someone else's definitions won't do. Telling me to jump out of a plane or fly into space will just make me feel anxious and dissatisfied. Then again, for the hardcore thrill seeker, driving through the countryside on a quiet Sunday afternoon might seem boring and wouldn't fulfil their sense of adventure either. It's an individual thing.

This might seem quite simplistic, but when you think about it, the 'actualisation' of a value will always mean something very different from one person to the next. Take the example of family. For some, living life in line with this value might mean visiting their family every single Sunday. For others, it might mean caring for elderly parents. Some people may practise their value by supporting their relatives financially. For others, it might mean maintaining close bonds whereby the emotional highs and lows are shared.

Let's consider the value of honesty. What does it really mean to be honest? Does honesty mean not lying? Admitting our mistakes? Being transparent and open about our actions and intentions? If so, how often and under what circumstances? Does it mean keeping promises? Or expressing ourselves in a genuine and authentic way? If so, to everyone or just some people? What about on social media? What about when someone asks how we are and we automatically say, 'Good, thanks' even though we're far from good? I don't have the answers to all these questions. Again, it comes down to what honesty means for you. This is why it's important to go through the process of defining each value so that we know for certain whether we're acting in alignment with it or not.

The bottom line is fulfilment comes when our values and our actions are in sync. This means making choices for ourselves every day that fulfil some or all of the values we hold dear. To do this, we need to build on the work of the previous chapters. Hopefully in Chapter 9 you were able to identify and narrow your list of values down to ten or less. In Chapter 11, you were able to assess your alignment with those values in your day-to-day life. Now consider these four key questions:

1. What is my value?
2. Am I, for the most part, living my life consistently with this value? What things might get in the way of that?
3. What current behaviours are consistent with me living life in line with this value?
4. What additional behaviours might help me live life more consistently with my values?

Rather than giving you hypothetical examples, I asked two of my clients to answer these questions. Here's what they had to say.

Christian, forty-three

1. *What is my value?*
Family

2. *Am I, for the most part, living my life consistently with this value? What things might get in the way of me living life consistently with this value?*
Mostly yes, but things that can get in the way include being busy at work, fatigue, and sometimes forgetting to prioritise family or even taking them for granted.

3. *What current behaviours are consistent with me living life in line with this value?*
Visit the family once a week for Sunday lunch; have regular contact via text and phone calls with parents; babysit my sister's kids

sometimes; cook dinner for the family sometimes; help out with things around the home that might need fixing.

4. What additional behaviours might help me live life more consistently with my values?
Help my father with fixing those things around the house that I said I would do but never did; put Grandma's recipes into a book and distribute it to all the children and grandchildren; organise a big family lunch for Mum's birthday this year.

Adam, twenty-nine

1. What is my value?
Creativity

2. Am I, for the most part, living my life consistently with this value? What things might get in the way of me living life consistently with this value?
Not really. Creativity doesn't feel like it fits naturally into my life at the moment. My life is quite structured and my work is analytical, which means that there are not many moments that feel like they easily allow for creativity. I'd need to actively schedule and block out time for creative pursuits.

3. What current behaviours are consistent with me living life in line with this value?
I have a sketch book with random drawings and doodles in it, but don't seriously do anything with this.

4. What additional behaviours might help me live life more consistently with my values?
Enrol in a class of some sort, perhaps cooking, pottery, photography or art and maybe even ask a friend to join me. I could buy

some paints and canvas, or even start with one of those 'Paint By Numbers' activities. I also have a guitar from when I was a child that I could start playing again. Leaving it out in the open would remind me to play and make it quick and easy to pick up some evenings after work.

•

As I was reflecting on what gets in the way of us living in line with our values, and what behaviours we might practise to remedy this, my good friend and neuroscientist Tanya Duckworth came to mind. Tanya and I have bonded in recent years over scientific research and life in academia. Tanya has faced some significant health problems, but still lives a rich and meaningful life. I spoke to her about how she manages to keep aligned with her values, despite her considerable health challenges. She was kind enough to share her thoughts and reflect on the four questions above.

Tanya, forty-four
1. *What is my value?*
Meaningful and interesting career

2. *Am I, for the most part, living my life consistently with this value? What things might get in the way of me living life consistently with this value?*
Yes, although dealing with endometriosis, PTSD and other health concerns frequently means that I am unable to function well at all.

3. *What current behaviours are consistent with me living life in line with this value?*
Flexibility and adaptability. I often have to make decisions and adapt to my health conditions. I make the most of the time when I'm

feeling well, or the couple of years following a surgery (maybe more if I'm lucky). During this time, I try to embrace life, working on my PhD, socialising, being outdoors and pursuing a creative activity like playing piano or guitar, or drawing or painting.

I work part-time so I can be flexible with my workdays. Additionally, I seek out work and study that give me the flexibility of working from home. I credit my resilience and perseverance to creative problem-solving – which is a big part of what my PhD is about. I don't think I would have continued to pursue my love of neuroscience and creativity if my overall purpose wasn't to use my personal experiences as someone living with PTSD to try in some way to improve the lives of others. I make the most of every spare second I have when I am symptom-free or symptom-minimal.

4. *What additional behaviours might help me live life more consistently with my values?*
I've been thinking a lot more about this lately as my disease has returned and my symptoms are worsening. I am not necessarily able to live the exact life I wanted for myself, travelling the world and living in different cities for research. I love what I do, but my body and the treatments (or lack of) currently available won't allow me to pursue exciting post-doctoral research overseas. Because of this, I need to look at what career alternatives are better options for me and still aligned with my values. So again, flexibility and problem-solving are important. I'm looking at scientific writing or running a private healthcare practice that I can do from home but that still involves helping people, and maybe helping people with endometriosis as well as PTSD. I can also stay connected with amazing researchers I meet via social media.

•

Values-aligned behaviours in and of themselves won't make us feel happy moment to moment. But they will contribute to a bigger and broader sense of satisfaction in life.

A number of years ago I saw a nineteen-year-old client named Ryan. He was living at home and had identified family as a core value. He wanted to do more to live life in alignment with this value so he decided to help out around the home. He set himself the task of doing the dishes more consistently. I have a vivid memory of him telling me, 'It doesn't make me happy and I certainly don't enjoy doing it. But I do feel better about myself and my relationship with my family for it. While I'm living at home, it's one of the ways I can contribute and show my gratitude.'

This is a good example of the difference between values-driven behaviours versus behaviours and activities that simply give us a dopamine hit. I doubt the dopamine was going off when Ryan was doing the dishes, but his overall sense of self and contentment in life improved.

It's easy to stay aligned with some values, but others require a bit more creative brainstorming. Take for example the value of flexibility. How on earth do we practise flexibility? Well, perhaps you set aside time each week to explore new perspectives or alternative solutions for current projects or tasks on your to-do list. Maybe you spend some time learning a new skill. Perhaps you change up your usual daily routine, or decide *not* to plan anything for the weekend and see what naturally unfolds. You could also do some 'mindfulness' practice (more on this later), which is ultimately all about helping us be flexible with how we relate to our thoughts, feelings and experiences.

In reflecting on what behaviours could help you live your life in alignment with your values, try to find both big *and* small actions. While it is great to fulfil the values of adventure and travel by booking a six-week overseas holiday, for most of us, this isn't something we can do very often, and we don't want to have to wait years to be able to fulfil this value! Sure, we can spend time on travel websites planning where to go and what to see, but are there other actions we can take to help us align ourselves with this value more consistently? Perhaps we take a road trip to a country town that's a couple of hours away over a long weekend. Perhaps we visit family or friends interstate at Christmas or Easter. Perhaps it's time to check out some local national parks and try camping. Ever heard of those tree-top adventures? They're basically obstacle courses made up of ropes and zip-lines, up high in the trees. They fit you with a harness and off you go! I've not yet been, but it's on my to-do list and could be a relatively easy way to live my value of adventure.

The value of achievement is important for a lot of people. I'm certainly looking forward to the sense of achievement I'll feel when I finally see my book on the bookshelves! But that's a big and rare milestone. In order to reflect on the value of achievement more regularly, I have a colleague who shares a prompt every Friday with a group of us, asking us to reflect on our 'weekly wins'. Those wins can be absolutely anything, but the prompt and reflection help us stay aligned with the value of achievement. I often encourage my clients to reflect on something they're proud of – the big achievements, but also the little ones. They all add up.

Do you remember Ben? The uni student who was depressed and dissatisfied and was escaping life by playing video games? Ben was training hard in long-distance running and studying accounting to take over his father's business one day. He'd stopped playing guitar since his bandmate moved overseas and instead spent a lot of time playing video games. Ben realised that he wasn't living life in a way that was consistent with his values. He was prioritising what his family valued – hard work, financial stability and sporting success. And this meant he was neglecting his personal values of creativity, freedom, fun and spontaneity.

My suggestion to Ben wasn't to drop out of university and book a one-way ticket to Alaska. In fact, I would have encouraged him *not* to make such an impulsive decision had he wanted to do that. Instead, I helped him figure out what living a life of creativity, freedom, fun and spontaneity might look like in the here and now. We workshopped how he could make small adjustments that would provide him with greater fulfilment, while also continuing with his studies and sport. Again, it's all about balance.

He decided to take a break for a few months from his long-distance running. He told me he could 'do with a few more sleep ins', as he often trained early in the mornings before heading to class. But he loved sport so he joined the uni basketball team which was both fun and a good way to socialise. He also planned a week-long surfing trip with a mate in the summer holidays on the Gold Coast. These small adjustments helped him stay connected to his love of outdoor sports, which was still a big part of his identity, but also aligned him with his values of fun and freedom.

Ben continued with his uni studies as he wasn't sure what else he wanted to do, but he had a new-found freedom to explore other options at the same time. He set up a meeting with a careers advisor and they discussed the different directions his studies could take him. He also dug out his guitar and reconnected with some old friends who were musicians. Every couple of weeks they'd get together for a jam at someone's place, and they talked about playing some gigs at local venues. These purposeful actions aligned with Ben's values of spontaneity and creativity.

Ben and I then discussed how he might reduce the amount of time he spent video gaming, which he was all for. We talked through some strategies which you can read more about in Part 3. However, we found that over time, he naturally had less of a desire to immerse himself in that world. He was happier and more satisfied in life and didn't need to constantly escape into alternate realities.

•

Hopefully you now have a better understanding of those core values which give you meaning and fulfilment and the actions required to ensure you're living them. The self-reflection exercises in the last few chapters provide the foundations for you to set some practical and behavioural goals. If you'd like to do some more self-reflection, you could try 'stream-of-consciousness' journalling, where you take a pen and paper and write down whatever comes into your mind. Don't worry about grammar, structure or tone! Just write. It can be a real surprise what you end up with.

How do you remember your values and make sure you're working towards them? You need to continually remind yourself of what they are. Having them written down on a piece of paper on your bedside table can help. Or maybe reminders in your phone will work best for you. Some people like to make a 'values vision board'. Or perhaps you could find an 'accountability buddy' with whom you share your values and regularly check in to see how you're travelling. I worked with a client once who would change her home screen on her phone every morning as a way of reminding her of her focus for that day.

Knowing your values is just the beginning. The next step is translating them into those actions that shape your daily life. Keep your values and action goals close by as we begin Part 3. We'll look at how we can moderate our dopamine-driven behaviours and keep on the path towards a values-driven life.

LET'S REFLECT

- Before reading this chapter, how much had you thought about the choices and actions you take and how values aligned they might be?

- Sometimes a goal can encompass more than one value. Do you have any actions or goals you'd like to fulfil that align with more than one value? If so, which ones?

- As you progress down the path of setting goals in line with your values, what practical steps can you take to ensure those goals become integral parts of your daily life, rather than distant aspirations?

- How might you remember the goals you've set for yourself, and remember to act in line with them? Are there any strategies or reminders that may be useful for you?

PART 3
BUILDING BEHAVIOURAL CONTROL

Congratulations, you've made it to Part 3 and are now hopefully ready to make some practical changes! In this final section of the book, I'll walk you through the steps that will help you take a break from your chosen dopamine-driven activity. When you do this, you're bound to experience urges or cravings. That's normal and to be expected. But don't worry; we'll discuss a range of psychological strategies that have been proven to assist with managing these uncomfortable sensations. We'll also look at how you can replace your dopamine-driven activity with values-aligned actions. At the very end, we'll consider the progress you've made and what steps to take next. And yes, you'll be faced with another choice! Keep your dopamine-driven activity on the shelf, or reintroduce it in a controlled way. The power will be in your hands.

Chapter 14

The power of plastic

In 1949, the Canadian psychiatrist and psychoanalyst, Donald Hebb, famously wrote, 'Neurons that fire together, wire together'.[1] It's a bit like the old saying 'birds of a feather flock together', but for your brain chemistry!

What Hebb meant is that a connection or pathway is formed in the brain whenever we practise a new behaviour. The more we repeat this behaviour, the more consolidated that pathway becomes. Every time we think in a certain way, do a particular task or feel a certain emotion, the pathway becomes stronger and stronger. It's what we refer to as 'Hebbian learning'.

The stronger a pathway is in the brain, the easier and quicker it is to perform that behaviour the next time. It's a bit like the experience of speeding down a six-lane highway compared to trekking through the wilderness. Driving down the highway is usually effortless and often fast. The highway represents a pathway in the brain that is well established. It might be made up of particular tasks that we repeat often, or patterns of thinking and specific emotions. As I'm typing right now, I don't need to search for each of the letters on the keyboard. I've done it so many times that my brain has formed very strong pathways for my fingers to know exactly where each letter is. The pathways in my brain that connect the spelling of words to my fingers are like a six-lane highway.

Now imagine doing a new task for the very first time. It'll probably feel less like driving down a highway and more like hiking through the wilderness! It's not easy and takes a considerable amount of effort. You've got to be careful not to trip over fallen logs, you have to watch out for snakes, and you need to carve your way through the bushland. The second time you trek through the wilderness, it'll be a little easier because you can follow the path you laid out the first time. Trek through it a few more times and you'd have a path that someone else could follow. Over time, the path becomes more established and easier to travel. Eventually, it might become so routinely used that it gets developed into a six-lane highway! The point is that the more we repeat a behaviour, the more consolidated the pathway in our brain becomes.

Another example is learning to drive a car. When you begin learning and are on your L's, there's a lot to consciously think about. Accelerator on the right, brake on the left. Handbrake off, indicate to turn, check mirrors, check blind spots, give way to your right (if you're in Australia) and lots more. But as you spend more time behind the wheel, the process becomes easier and requires less conscious thought. Driving starts to feel automatic and we do the checks and the accelerating and braking without even (seemingly) having to think about it. This is because our brain has formed a pathway – known as a 'neural network' – related to driving, which has become consolidated with repetition. The more consolidated a neural network becomes, the quicker it fires, and the easier it is for us to access and action it.

Now, what does all this have to do with changing our dopamine-driven behaviours?

Well, the bad news is that the more times you have sat on your bed watching Netflix, the more developed the pathway is in your brain that associates 'bed' with 'movies'. I have a six-lane highway that connects opening my phone with clicking on the Instagram icon. In fact, the neural network is so well developed, that even if I delete the app, I still unlock my phone and instinctively go to press the icon!

The more you've performed the action you want to change, the more developed and automatic the pathway will have become in your brain. Combine the principles of Hebbian learning with what we already know about the drive that dopamine gives us to engage in certain behaviours and you can see why these patterns are so hard to shift!

But here's the good news: the human brain can *change*. It's not a static organ, it's dynamic. The phenomenon is what we refer to as 'neuroplasticity' (the word 'plastic' coming from the Latin word *plasticus* and the Greek term *plastikos*, meaning 'moulded or formed'). Neurons are the building blocks of the brain and they can change and adjust. In the same way that they worked to form the six-lane, phone-to-Instagram highway, they can also form new, healthier pathways.

While neuroplasticity means the brain can change for good, it can also change to develop an addiction. With most addictions, the brain builds up tolerance and physical dependence, meaning the user needs more and more to achieve the same effect. This can happen to anyone who repeatedly uses drugs like sedatives, opioids, cocaine and nicotine.[2] Among the severely addicted, the changes in the brain can be very long-lasting.[3] And the changes to dopamine release caused by the drugs tend to be greater than that of naturally occurring rewards, meaning that drugs can easily overpower our natural or useful behaviours.

Happily, we can also utilise the power of neuroplasticity to do remarkable things. In a well-known research study, the Irish neuro-scientist Eleanor Maguire examined the brains of British taxi drivers and found that specific use of a particular cognitive process changed the shape and structure of their brains.[4] She looked at the drivers' 'visuospatial memory', that is, the cognitive process required to remember how to get from one location to another and found that their abilities were almost superhuman! The drivers have to do a lot of training, commonly referred to as 'The Knowledge'. Over time, they manage to learn approximately 25,000 streets in London, thousands of landmarks around the city, and how to navigate *without* the aid of technology-assisted navigation. MRI studies have shown that their posterior hippocampus region is larger compared to other people.

But the brain can do more than just learn new languages and maps of cities. The brain's tissue has the incredible capacity to take on new functions.[5] For example, for some people who have experienced brain injuries, neuroplasticity can result in the undamaged regions taking over functions like sensory or motor capacities. In *The Brain that Changes Itself*, the British psychiatrist Norman Doidge writes about people who recovered the use of paralysed body parts, deaf people who learnt to hear, and people who found relief from pain using exercises to retrain their neural pathways.[6]

Before Mora Leeb was born in September 2007, she suffered a massive stroke in utero, killing most of the cells in the left hemisphere of her brain and leaving her with a rare neurological condition. At just nine months old, surgeons decided to remove the damaged left-hand section of her brain. This is the part that plays a critical role in producing and understanding speech and language. You'd think this would have been catastrophic. But at age sixteen in 2024, Mora is very much your typical teen. Although she processes speech slowly, she can communicate, tell jokes and understand language. Thanks to neuroplasticity, her brain's right hemisphere was able to take on some of the functions usually done by the left. What's more, she plays tennis, can do Sudoku puzzles, loves to sing and can even ski![7]

The procedure Mora underwent is known as 'hemispherectomy'. Dr Dorit Kleimann, a professor of psychology and neuroscience in the US, did a study in 2019 looking at the brain functioning of six young people who had undergone hemispherectomies before the age of eleven.[8] The functional magnetic resonance imaging (fMRI) data showed that despite missing half of their brain, they all had the same major networks as in a healthy brain with two hemispheres!

If Mora and others who have had half of their brain removed are not just living but *thriving* thanks to the power of neuroplasticity, you and I can surely make use of the brain's incredible capability to build and strengthen new pathways, while weakening old ones. We'll now learn how to harness the brain's neuroplasticity and change those pathways that have been reinforced by dopamine. It'll be a two-pronged approach – we'll look to alter neural pathways and reset dopamine levels on a neurochemical level, while also building behavioural control over any unhealthy habits.

LET'S REFLECT

- What are some examples of 'neurons that fire together, wire together' in your life?

- What behaviours have you developed that have become automatic? For example, walking, riding a bike, driving, typing, writing and speaking another language.

- How did it feel when you first started doing that behaviour? Was it hard? What helped it become quicker and more automatic over time?

- Consider how understanding neuroplasticity and habit formation might influence your future goals. How might this information help you foster resilience, adaptability and continued growth?

Chapter 15

Take a break

It is said that insanity is doing the same thing over and over again and expecting different results. If we want different outcomes, we need to do something differently! And this means more than just thinking about how nice it would be if things were to change. We need to put strategies in place to ensure it happens.

This is where we start making practical changes.

I need to reiterate that these are not recommendations for people who have severe addictions to substances like alcohol or drugs. Withdrawals and cravings need to be carefully managed by a health professional. The changes we are about to discuss are for the everyday person who has some unhealthy habits or patterns over which they'd like to gain greater control. And the reality is many of the behaviours we might target, such as overeating or spending too much time on our phones, aren't things we can or would want to completely get rid of anyway!

I live by the phrase 'nothing changes if nothing changes'. I say this not only to my clients, but also to myself every time I find I'm wishing something were different. *If we want something to be different, then we need to change it.* I hope you are ready to change things for the better.

Back in Chapter 7, you chose a target behaviour. Keep this target behaviour in mind as we work through the following strategies. You can apply these strategies to a different target behaviour later, though I'd recommend focusing on one thing at a time and keeping it straightforward. When change feels too big or overwhelming, it's less likely to be successful and we can very quickly slip back into old habits. Change is more likely to become an automatic habit when it's simple.[1] It needs to feel easy to integrate and sustain within our everyday lives.

But first thing's first – let's start with taking a break.

Let me paint a picture of a common situation that unfolds in the therapy room. I'll use Joe as an example. Joe contacted me at a time when my books were full. I recommended another psychologist, but he wanted to wait until I had availability. When I finally saw him, he told me about an unhealthy reliance he'd formed on pornography. He was distraught. The behaviour felt out of control. His pornography use had begun to affect his relationship and his partner had told him to seek help. It made it hard to focus at work and be present with his family. When I first met him and he told me his story, I could tell that he was genuine. This behaviour was having a really negative effect on his emotional state and peace of mind. I suggested that his first step should be to 'take a break' and have a period of abstinence. I explained that this would be challenging and that he'd experience urges and uncomfortable emotions, but that I could help him learn skills and strategies to manage the discomfort. I told him that change wasn't easy, but it was worth it.

Joe, like many clients faced with such a recommendation, hated it. Even when someone knows how badly their behaviour is affecting them and their loved ones, the suggestion of taking a break is often met with hesitation, if not resistance. This is usually followed by a list of reasons (and excuses) as to why taking a break wouldn't be possible, or why it's a terrible idea. Here are just some of the excuses I've heard over the years:

- Research says that people who drink one to two glasses of red wine a day live healthier and happier lives. Therefore, I don't think I should totally stop drinking.
- My friend has been using mushrooms and said they had a major trip which helped them process their trauma about their parents. I think there might be benefits to these drugs.
- My parents have accepted that I'm a stoner and have given up pushing me to get work. If I stop using, then they'll expect more from me.
- I saw a guy on TikTok who said that weed cured his ADHD and OCD.
- I saw a news article that said that masturbating is actually good for your mental health because it releases endorphins and makes you happy.

Sometimes people come to therapy thinking that I have the magical ability to 'talk them better'. That there's something I'll say that they've never heard before, that will change how they feel and how they interact with the world. Suddenly they'll no longer feel the urge to watch pornography excessively, binge-drink alcohol or coffee, or constantly post on social media for external validation. If only it worked like that! Sure, I like to think that my clients leave with some meaningful takeaway messages. But what people are often hoping for, which isn't possible, is for me to take away a *feeling*. To take away the urge to watch porn. To take away the discomfort of sitting with certain emotions. To take away the stress of work, the tension in a relationship, or the heartache of a breakup. Unfortunately, this is not something I can do.

However, what I *can* do is equip them to be able to feel differently about themselves and make changes. But that requires a person to be ready and willing to do the work.

Why is this relevant? Well, if you've got this far in the book, congratulations! You're the equivalent of the person who has had several therapy sessions with me. The person who has told me their story, shared their struggles and explained how they'd like to feel better. We've talked and considered the important things in their life and how they want to live.

Now it's time for you to take action and actually make those changes.

You have a choice.

You can continue to read on as an observer on the sidelines, experiencing what it would be like to be a spectator in the therapy room, but not officially a part of the action. You can consider what it might be like to implement the strategies we'll talk about. How your day-to-day life might change, or not change. If that's you, then I commend you for reading on and considering these options. However, I also say, don't get to the end of the book and expect for things to be all that different! Nothing changes if nothing changes, and if you don't want to implement all these strategies, then that is fine. Just have realistic expectations for how life will look at the end of it. I would, however, encourage you to select one – just one – small change that you can consciously and purposefully implement.

But if you are ready, willing and able to take action, my first recommendation is to take a break from your target behaviour. And by 'take a break' I mean stop it. Abstain from it.

If you like, treat taking a break like an experiment. Give it a go for say, two weeks, and see what happens. There's no lock-in contract that says once you give the behaviour up you can never go back to it. At the end of the two weeks, you control what you do next. But we won't find out what the relationship with the substance, device or behaviour is truly like until we give ourselves some distance from it.

So why is taking a break important? Well, it all comes back to dopamine and changes in our brain chemistry. We already debunked the 'dopamine detox' myth in Chapter 2 so it's definitely not that! Abstaining from a substance or activity for a period of time allows the brain to readjust its baseline level of dopamine and its sensitivity. When you consistently engage in an activity that stimulates dopamine, the brain adjusts itself to regulate its state of homeostasis. The goal is to feel *less* dominated by unhealthy stimuli.

But there's more to taking a break than just dopamine. I'm a scientist, so I like data. I appreciate the precision and clarity a data set can provide. I like that data can help us measure change, but also make meaning of those changes. Taking a break and viewing it as an experiment is useful because you collect data about yourself. Treating taking a break like an experiment is a good approach if you're not sure of the exact effect something is having on you. For example, you may have a suspicion that when you go out with your friends on the weekends you drink a bit too much. It's not a big problem but cutting back on the money you spend on alcohol probably wouldn't hurt. You're not turning up late for work, cancelling social plans, developing liver disease or fighting with your partner about how much you drink. But you've considered drinking less. A period of abstinence will help show us the real cause and effect between alcohol and how we might be feeling. You'll already have the data on how going out and drinking affects you. If this were an experiment, you've probably repeated it many times over and found similar results each time.

Now let's collect data on the *opposite* experiment.

Set yourself the goal of *not* drinking alcohol for a period of time and take note of the differences. Imagine you're a scientist. Approach it with curiosity. How do you feel? Do you have fun? How do you feel the next day? What are you missing? What are you gaining? Approach it with genuine, open curiosity, so that when the period of abstinence ends, you can make an informed decision as to how you'd like to proceed.

Taking a break is as much about learning to build behavioural control as it is about making neurochemical changes. While we aren't engaging with the target behaviour, we're bound to experience some sort of urge or craving. Just like when my phone sits on my desk while I'm working and I feel the urge to pick it up and check Instagram, you'll feel the urge to do whatever it is you've told yourself you won't do. This is good! This is *information*. Feelings and experiences are information. Try not to label them as 'good' or 'bad', just note them as 'data points'.

Now, the next (perhaps predictable) question I'm asked, once people are okay with taking a break is, 'How long for?'. How long is needed in order to build the behavioural control and reset homeostasis in the brain? That's a tough one to answer. It depends on a number of factors, but how 'hooked' you are to the substance or behaviour is critical. It also depends on how much and how frequently you use the substance or engage in the activity, as well as your age, your biology, your genetics, and your level of tolerance.

There are all sorts of theories going around. In Silicon Valley, you'll find executives and entrepreneurs happy to lock themselves in a vacuum-sealed box for a 24-hour period. They deprive themselves of all sources of joy or stimulation for this 'dopamine reset'. No technology, no food, no phone, no music, no talking to people. Nothing that might spark joy and release dopamine. That's quite an extreme approach if you ask me, and it's unfortunately probably not the silver bullet they're looking for. It's unsustainable, unrealistic and not likely to lead to long-term change. And of course it's not scientifically valid, as we can never fully stop the release of dopamine, even in a vacuum-sealed box!

But let's not confuse that with something that will have a longer-term effect. Dr Anna Lembke in her book *Dopamine Nation* recommends abstaining from our target behaviour for one month to achieve a 'dopamine reset' and build enough awareness to notice patterns of thought and emotions that may arise.[2]

At many rehabilitation centres or in twelve-step recovery programs, you'll hear 'ninety days' talked about a lot. Reaching that ninety-day abstinence milestone is celebrated and is quite an achievement for those who have struggled with addiction. In fact, if you make it to ninety days, you'll likely to be given a 'chip' in recognition of your success. These programs recommend a period of abstinence for at least ninety days before making further decisions.

There is evidence to suggest that with serious addiction, say drugs or alcohol, the brain's neuroplasticity takes quite a bit of time to reset. Brain imaging studies show rapid activation of the brain's reward pathways still occurs when drug-related cues are displayed to people who haven't used drugs for a month.[3] Even though they've stopped taking the drug, the activation still happens, evidenced by an increase in blood flow to the reward pathways. This occurs even when they're shown cues for a tiny *thirty-three milliseconds* – cues so brief that they don't even reach our conscious awareness.[4] And when this happened, they reported cravings, with the strength of the craving directly related to the amount of dopamine released.[5]

Research does not tell us definitively how long it takes to rewire the state of homeostasis in the brain. My recommendations, based on studies combined with clinical experience, are as follows. If I'm seeing you in my clinic for a serious addiction, I agree with the recommended period of a minimum *ninety days* of abstinence without relapse. This is a good amount of time to aim for. Historically, it was thought that *total* abstinence during the ninety days was the only option for people with serious addictions. More recently, research has suggested that it's enough for people to just *reduce* their consumption.[6] However, in my experience, if the addiction is severe, merely reducing the amount of use is risky.

If you're not dealing with a serious addiction, choose a period of time that feels manageable to you. It could be two weeks or a month. Don't go for the '24-hour or 48-hour fast'. That's not enough time for us to work on anything within ourselves other than perhaps fighting off the occasional urge and distracting ourselves through the discomfort. Aim for two to four weeks to start with.

Here's another important to thing to remember. If simply abstaining from your target behaviour or substance is what makes you feel better, then congratulations! You've struck gold. For many of us, however, it's not that simple. Historically, clinicians considered the process of detoxification to be the treatment for drug addiction, but we now know that that is far from the case, with detox being only one part of removing the drug from the body.[7]

Both animal and human research studies tell us that cravings gradually increase in the weeks and months following the initial period of abstinence.[8] This is a process known as 'incubation' – the progressive increase of cravings during early abstinence.[9] These cravings can remain intense for extended periods of time, in many cases for up to six months after the period of abstinence has begun.

Rats who have been trained to self-administer cocaine have cravings which are significantly stronger after one to two months of abstinence, compared to one to seven days.[10] This also applies to heroin, nicotine, alcohol, methamphetamines and opioids. Studies have shown the same process also occurs with food. Food cravings during the incubation process have been seen in rat studies for standard food, as well as high-fat and high-sugar foods.[11] And it's not just rats. The incubation of cravings has also been observed in humans (studies have focused on cigarette smoking, methamphetamine use, alcohol and cocaine).[12] Incubation of cravings is also said to be present in addictive behaviours like gambling.[13]

This is all to say, that 'taking a break' often involves a lot more than simply 'taking a break'. There is no magic amount of time. While most of us won't be dealing with severe addictions and won't experience severe withdrawal symptoms, it's still going to be unpleasant. What's really important when we take a break from smoking, binge-watching TV or eating too many sugary foods, for example, is what *else* we do during that time. What we learn about our thoughts, our emotions and internal experiences and how we reflect on our urges and desires for certain behaviours. Do we replace smoking cigarettes with vaping? Do we replace alcohol with extra food? Do we replace one digital pursuit with another? It's at this point where the real, long-lasting change takes effect. This is where we get to reflect on our values and start doing things that are aligned with those values. Taking a break allows us to build behavioural control, but it also allows us to experience the feelings that would have been masked or avoided, and to do things that are meaningful and powerful.

I hope you've now decided take a break from your target behaviour and have locked in an amount of time. We'll next look at how to sit with the discomfort that'll inevitably arise, how to manage the urges to go back to the target behaviour, and how to fill that space with more meaningful pursuits.

LET'S REFLECT

- Change can be hard. What are some examples of changes that you have successfully made in your life?

- Have you been able to sustain these changes? If so, what helped?

- Have there been changes you've tried to make that haven't been as successful? What got in the way?

- Were the barriers to long-lasting change internal (e.g. your own thoughts and beliefs), or were they external (e.g. not having the resources to properly make the change)?

Chapter 16

Sitting with discomfort

I considered starting this chapter with a profound quote about stepping into the uncomfortable, perhaps something along the lines of how 'growth happens when you're outside your comfort zone' or 'change begins at the end of your comfort zone' . . . but I think it's more realistic to start with this: *sitting with discomfort is hard*. There's no getting around that. Why? Well, we all have an innate inclination to *avoid* unpleasant feelings.

In the late 1800s and early 1900s, the founder of psychoanalysis, Sigmund Freud, put forward what's known as the 'pain-pleasure principle'. Freud thought that people make choices in life which either decrease pain or increase pleasure, and that the intrinsic motivation to do so satisfies both biological (or physical) and psychological (or mental) needs.[1]

While physical pain and mental pain are different, research suggests they share some neurological similarities.[2] 'Social pain', for example, rejection, exclusion or loss, activates the brain's physical pain-related neural pathways. And people who are more sensitive to physical pain are often more sensitive to emotional pain as well. The way we describe rejection, for example, is consistent with feelings of physical pain, such as *hurt* feelings or *broken* hearts. In one study, people who had recently experienced an unwanted romantic relationship breakup were asked to complete two tasks.[3] In the first task, they had to look at a photo of the person who broke up with them and think back to that experience of rejection. In the second task, they received a painful heat stimulation from a thermode (a small metal plate) attached to the skin. Results showed that the pain-related pathways in the brain (specifically, the dorsal anterior cingulate cortex, anterior insula and posterior insula) responded to *both* reliving the rejection, as well as the painful heat stimulation. Physical pain as well as emotional pain activate common neural regions, even to the point where parts of the brain are not only responsive to actual social rejection, but they also respond to the mere *possibility* of social rejection.[4]

The 'relatability' of emotional pain is why we have so many songs, movies, plays and artworks that depict heartbreak, loss, grief and sadness. Humans have been making art which portrays misery and despair forever. The ancient Greeks were known for staging 'tragedies', a popular form of drama. Vincent van Gogh's painting *Sorrowing Old Man* depicts someone who is despairing emotionally and also sick. The Portuguese 'fado' and the Irish 'lament' are both traditional forms of music that capture expressions of sadness and pain. And in more modern times, countless chart-topping songs have been written about love lost, from Toni Braxton's 'Un-Break My Heart' and Adele's 'Someone Like You', to Gloria Gaynor's 'I Will Survive' and 'Nothing Compares 2 U' by Sinéad O'Connor. Emotional pain that people feel physically doesn't just exist, but is deeply relatable and an intrinsic part of the human experience. We *all* feel emotional pain and discomfort.

Of course, it's natural to want to avoid such feelings. After all, who wouldn't want to feel good all of the time? As we've already discussed, much of modern life has been curated to help us feel fab and to do so quickly and easily. However, there are benefits to mastering the ability to sit with emotional pain and discomfort. But what does it actually mean?

Sitting with discomfort is essentially allowing ourselves to feel unpleasant emotions and experiences. This might be an unpleasant feeling such as sadness, anger or grief. It might be an uncomfortable physical sensation such as butterflies in the stomach when we're feeling nervous. Or it might be a strong urge or craving for something like a cigarette or a glass of wine. Instead of pushing the feeling away, distracting ourselves with another task, or avoiding the feeling all together, we must do the opposite. We must make space for the discomfort. Allow it to be there and practise being 'with' it. This is a key part of our ability to manage and regulate our emotional experiences.

Being able to do this first requires us to be aware of what these feelings actually are. This is a skill that psychologists refer to as 'interoceptive awareness', that is, the ability to be aware of internal sensations and cues such as our heart rate, our hunger level, our fullness, our experience of pain, our urges and our emotional sensations. The awareness of our internal experiences and states allows us to regulate our emotions and effectively manage and respond to life's continual ups and downs.

In order to effectively regulate our emotions, we have to be able to accurately detect and evaluate our internal cues. This is central to managing social interactions, maintaining healthy psychological wellbeing, and preventing some forms of emotional distress and mental ill health.[5] It also helps us deal with highly stressful situations more easily, improves the quality of our relationships with others, and allows us to respond to difficult interactions with greater composure and control.

People who struggle to regulate their emotions are at increased risk of developing a wide range of mental health concerns in both adolescence and adulthood.[6] Researchers Amelia Aldao, Susan Nolen-Hoeksema and Susanne Schweizer conducted a large-scale review that showed evidence of the links between emotional dys-regulation and depression, anxiety, eating disorders and substance abuse.[7]

We begin learning how to regulate our emotions right from the moment we are born. These skills are developed through the way our parents and caregivers respond to us when we are crying or dis-tressed. We also learn by watching and modelling the behaviour of other adults around us. If we see adults resolving conflict through calm communication, we're more likely to adopt this method our-selves. However, if we see them managing their emotions through loud arguments or heavy drinking, we'll learn this way of coping too. And while arguing and drinking may regulate emotions in the short term, they're definitely not the most *effective* ways to go about it!

So how and why is this relevant to dopamine? Well, as you go through the period of abstinence that you set for yourself in the previous chapter, you're going to feel things. There's no getting away from it. The basic building blocks of cognitive and behavioural therapies are this: our thoughts, our feelings and our behaviours are all connected. The way I think about something affects how I feel, which then affects what I do. For example, if I walk down the street and see a dog and think to myself, 'That dog looks mean and scary and might bite me', I'm going to feel fear (emotion) and choose to walk on the other side of the street (behaviour). But if I see the dog and think, 'I love dogs, it's so cute!', I'll feel joy (emotion) and go up to it and pat it (behaviour). This is a very basic example of the links between these three components of our experiences. As our thoughts, feelings and behaviours are all linked, changing one has a flow-on effect on the others. The change you are making by taking a break (abstaining) is a behavioural one. And this behavioural change will result in changed thoughts as well as altered emotional states.

So there's no escaping it: taking a break from something that has been reinforced over a long period of time by dopamine will bring about uncomfortable feelings and sensations. The ability to notice these experiences and 'sit with the discomfort' strengthens our ability to regulate our emotions. But there is something else important to note. Remember how I said I love data? Well, internal experiences are data. They are sources of information. People are often quick to label pleasant emotions as 'good' and unpleasant emotions as 'bad'. However, *all* emotions provide us with information. Happiness might feel 'good' and anxiety might feel 'bad', but they're equally as important when it comes to understanding ourselves.

From an evolutionary perspective, human emotions are 'adaptive' and contribute to our survival. Emotions exist for a reason and they all serve a function. By taking away distractions and taking a break from your unhealthy, dopamine-driven habits, you allow your brain and body to receive more information in the form of emotional cues. Emotions are complex and play a crucial role in various aspects of human behaviour and social interaction.

Take the feeling of fear or anxiety, for example. Fear and anxiety are uncomfortable and often unpleasant. When people feel fearful or anxious, they often try to find ways to *stop* those feelings. How many times have we heard people try to comfort someone who is stressed by saying, 'Don't worry, you'll be fine!'. We don't like our own feelings of fear, anxiety or worry and we don't want other people to feel that way either. So, we tell them not to. However, fear serves an important biological function. It's a signal that warns us of potential or impending danger. Fear is a response to threats in our environment. It triggers the 'fight or flight' response and helps prepare our body to deal with potentially dangerous situations. So from an evolutionary standpoint, fear is very useful. Without it, our ancestors would have been eaten by tigers or fallen off cliffs!

Outside of my work, I have a background as a concert pianist and often get nervous before a performance. Rather than interpreting these nerves negatively because they don't feel good, I have to tell myself that they serve a purpose. Nerves tell me that what I'm about to do is important. I care about the performance I'm about to give and I want the audience to enjoy it. Reframing what the nerves mean helps me to channel the emotional experience in a useful way, rather than trying to avoid it (which is close to impossible!).

Anger is another uncomfortable emotion that many people try to avoid, but just like fear, it too serves a function. Anger tells me that I perceive there to be an injustice, that either I or someone else has been wronged. Anger provides us with this information but can also motivate us to make a situation better or advocate for change.

Sadness, on the other hand, is an emotion that has evolved as a response to loss. While unpleasant, sadness tells us that something or someone was important to us. Displaying sadness can also be a signal to others that we need comfort or support. For example, if we see someone crying, most of us have a natural urge to ask if they are okay or give them a hug.

Guilt is another unpleasant emotion that people naturally try to avoid, but guilt provides useful information as to our moral sense of right and wrong. Guilt tells us that we have done something that conflicts with our moral compass and can motivate us to apologise or make amends.

In this way, all unpleasant feelings serve a function and provide us with data. It's our job to recognise these emotions and make sense of the information they are communicating to us.

It's important to note here that some people regularly experience *disproportionate* emotions. That is, their emotions are more intense than the situation warrants, or they are not aligned to the situation. If this is the case, then it's crucial to see a mental health professional. Experiencing excessive anxiety in ways that aren't functional could in fact indicate an anxiety disorder and require treatment.

Now don't get me wrong. I don't want people to be sitting in a world of pain and suffering 24/7! In fact, the ability to be able to distract ourselves and put unpleasant emotions to one side is also useful when it comes to self-regulation (more on this in Chapter 18). Sometimes we have important things that need to be done and it's just not possible for us to be sitting with overwhelming emotions. If I wake up in the morning and need to get to a job interview, but I'm feeling really sad, I'll distract myself rather than risk becoming overwhelmed and missing the interview. The problems only start if we chronically avoid and push aside our negative or unpleasant emotions and never allow ourselves to sit with them. In fact, if we do that, these emotions will eventually come out more intensely and at really inconvenient times! There's never a 'great time' to sit with uncomfortable emotions, but practising it when we can will help us build this skill.

Embracing discomfort can be powerful. There's some truth to the saying about how growth happens *outside* of our comfort zone. When I think back to significant moments in history, or big changes in society, they often haven't come about by people being comfortable. I find it hard to imagine that Rosa Parks or the fellow travellers on the segregated bus in Alabama back in 1943 felt comfortable when she refused to give up her seat to a white man. And yet her act sparked the 'Montgomery Bus Boycott', a pivotal event in the American civil rights movement. In 2012, Pakistani teenager, Malala Yousafzai, was targeted and shot for advocating for girls' education. Despite this, she survived and continued her activism, becoming the youngest ever Nobel Peace Prize laureate. Continuing to do this work after being shot would have brought about a whole range of uncomfortable emotions, and yet Malala prevailed. There are many more examples like those of Rosa Parks and Malala Yousafzai.

So don't fear discomfort. As you go through the process of changing your relationship with dopamine-driven activities, you're going to feel uncomfortable. Modern-day society has developed in a way that promotes seeking out pleasure and avoiding pain. However, from an evolutionary perspective, pain and uncomfortable experiences are not bad things. They are unavoidable and are important sources of information.

As you take a break from your target substance or behaviour and start to shift the patterns of dopamine firing your brain, notice how you feel. Expect there to be times of discomfort. Notice the emotions and sensations in your body. Allow yourself to feel them. Remember that they are sources of information for you. They tell you about your feelings and emotional states; the good, the bad and the ugly. If we try to push these feelings or urges away, we'll end up replacing one unhealthy action with another. So many clients have told me that when they tried to stop smoking, they ended up eating more. Or when they cut out gambling, their drinking increased. Or when they stopped using social media, their TV binge-watching went up. This happens because they didn't allow themselves to feel the stress, the boredom, or the irritability; they didn't allow themselves to sit with it. Instead, they replaced one behaviour with another. Put simply, the more in tune we are with our own experiences, the better position we are in to make useful and effective changes.

LET'S REFLECT

- Growth is uncomfortable! What is something you've been able to do because you allowed yourself to feel uncomfortable?

- Regulating our emotions is important for our day-to-day functioning. What go-to strategies do you have for managing and processing your internal experiences?

- We all have different natural degrees of awareness of our internal state. Are you in tune with your inner states and experiences, or could you improve noticing how you're feeling?

- Sometimes when we push emotions away, they come out when we least expect them to. Have you ever experienced being overwhelmed by emotions at an inconvenient time because you hadn't allowed space for them earlier?

Chapter 17

Riding the wave

Now that I have (hopefully) convinced you that embracing discomfort can actually be beneficial, let's talk about how to do it effectively. How do we skilfully tolerate and even embrace the uncomfortable feelings that will inevitably occur when we take a break from our dopamine-driven activities?

Most people have heard of the term 'mindfulness'. It's by no means a new concept. In fact, it's ancient, with certain branches originating in Buddhist traditions in the East thousands of years ago. Yoga, which came from India, incorporates mindfulness and meditation, while closer to home, Indigenous Australians from the Daly River region in the Northern Territory used a mindfulness practice known as *dadirri*, which means 'inner deep listening' and 'quiet still awareness'.[1]

Even though most people have heard of mindfulness, their understanding of it varies greatly. Whenever I teach clients mindfulness exercises, or run training seminars for mental health staff on psychological principles, I always start off by asking what they think it means. Most say it's about 'not thinking' or 'emptying the brain'. Or 'feeling good', or a 'relaxation strategy'.

At its core, mindfulness is about being fully aware of the present moment without judgement. Dr Jon Kabat-Zinn, one of the medical professors who brought mindfulness to the West, defines it as, 'Paying attention in a particular way: on purpose, in the present moment, and non-judgementally'.[2] Our mind will wander – that's what the mind does. It will think about the past and the future. It will get distracted in thoughts unintentionally. Mindfulness allows us to notice where our thoughts have gone and gently guide them back to the present moment. The here and now.

The goal is *not* to feel good or calm and relaxed! These are certainly some of the benefits, but it's definitely *not* what we should be striving for. Nor is mindfulness about emptying the mind or being absent of any thoughts. This is practically impossible! The brain is designed to think. Even the Buddhist monks who practise mindfulness every day can't not think thoughts!

Unfortunately, these common misconceptions about what mindfulness is (or isn't) put up barriers to people engaging with it. If people try to practise mindfulness and can't empty their mind of thoughts, or don't feel calm and relaxed at the end, they'll tell me that it's 'too hard' and they 'can't do it', and usually give up. The truth is, these experiences, good and bad, are all part of it.

Let's do a quick exercise. This is something you can do to practise mindfulness when you are taking your break from your chosen dopamine-driven behaviour. It'll help you ride the discomfort. Stop for a moment and find an object nearby that you can easily reach. It doesn't have to be an interesting object. In fact, the more mundane, the better. Now, use your five senses to *notice* that object. What shape is it? Feel the weight of it in your hand. How heavy or light is it? Does it feel dense and solid, or empty and hollow? What's the temperature of the object? What colour is it? Are there multiple colours or just one? Are there any patterns, lines or textures? Are the colours vibrant and bright, or dull and muted? Are there any blemishes or imperfections? Notice how it feels on every fingertip. If you hold it up to your ear, does it make any sound? Does the sound change at all when you move it in different ways? Does the object have any discernible scent?

This is an example of mindfulness. It's simply the practice of noticing without judgement; of being present in the moment with whatever you have chosen as the focus of your attention. It's the art of using the senses to observe and describe and experience. Now, instead of an object, imagine that the focus of your attention is something *internal*. A feeling inside you. Use the exact same principles to notice, observe and describe that inner experience. What can you feel within your body? Are there any sensations? Are your muscles tight and constricted or relaxed? Is there any tingling? Any tension? Any aches or pains? Any discomfort? Where do you feel these in your body? Do they move through your body, or do they stay in the one place? Are the sensations dull or strong? Try to remain curious about the sensations. Notice if there is any resistance to or acceptance of the sensations. Can you cultivate a sense of openness and non-judgement? If you feel the desire to move, resist or avoid certain sensations, notice this urge too. Identify what emotion you might be experiencing. You might be feeling more than one. You might in fact be experiencing two seemingly conflicting emotions (for example, anxious *and* excited). That's okay. Just notice what they are. Again, notice if there is a resistance to experiencing these emotions, or if you can accept them.

Taking a non-judgemental, curious approach to our experiences allows us to sit with emotions and regulate them. It helps us to experience a range of emotions without feeling the need to avoid them or distract ourselves. This will be incredibly important and useful as you take a break from your target substance or behaviour. When feelings of discomfort arise, mindfully sitting with and embracing the moment will help you tolerate them and build your capacity to manage them when they arise again in the future.

Dr Daniel Siegel, a clinical professor of psychiatry at UCLA in the US, coined the saying, 'name it to tame it'.[3] This means that simply the act of naming or labelling an emotional experience as it is happening can reduce its distress and intensity. How does this work? Well, the part of the brain that is responsible for emotions, the limbic area, is a very old and primal circuit. It sometimes has us overreacting to situations that don't warrant a big emotional reaction, flooding our bodies with stress hormones such as adrenaline and cortisol, and causing us to spiral into catastrophic ways of thinking. By naming the emotion we are experiencing, we activate the prefrontal cortex, the part of the brain that is responsible for our higher order cognitive functions like reasoning, problem-solving, planning and impulse control. Naming our emotion therefore helps to reduce its intensity.

People who regularly practise mindfulness experience significant psychological benefits. Mindfulness can help reduce anxiety, depression, substance abuse and chronic pain.[4] Its effects can even be seen structurally within the brain. We've already talked about how the brain is plastic and can change. Well, practising mindfulness repeatedly is a great example of that. Functional magnetic resonance imaging (fMRI) studies show that mindfulness increases the grey matter volume and concentration within certain areas such as the left hippocampus and the posterior cingulate cortex.[5, 6] The increases in grey matter are related to improved learning and memory, emotion regulation, self-regulation and the acquisition of perspective – all highly useful psychological skills!

There's also a lot of research that shows that mindfulness-based interventions are very useful in helping people to manage pain. In cases of acute pain, mindfulness can have a beneficial effect on pain tolerance and pain threshold.[7] For people suffering chronic pain, for example back pain, mindfulness can help reduce the intensity.[8] It can help us manage our psychological discomfort, as our thought processes can make an already unpleasant or distressing situation feel even worse. Mindfulness helps us to reduce and remove necessary suffering that we create via our own internal experience, and just focus on managing the raw pain in and of itself.

Take a client of mine, Molly. Molly did not come to see me because she had an addiction. Rather, she was struggling to manage her emotions. Tragically, Molly had suffered trauma, abuse and neglect by her family and was stuck in a loop of suffering. She kept asking herself 'why?'. 'Why did this happen to me? Why did people treat me so badly? Why could I not protect myself? Why did people not care more about me? Why did others not step in and help me? Why does this still affect me?' These were all legitimate questions. But they were questions she'd asked herself for many years, and had never managed to answer.

Indeed, there were no answers that we could find. We could hypothesise, but doing so often just fed her loop of suffering and brought up more and more questions. Even when we were able to speculate about why the abuse had happened, the questions kept coming back. Unintentionally, Molly was creating excess suffering for herself and her neuroplasticity was reinforcing this. We worked together to validate how distressing her experiences were, while also removing the additional layering of suffering to try to stop the spiral of 'why' and deal only with the raw emotional pain itself. Life is painful. For some it is more painful than for others. But we can all get 'stuck' adding additional suffering to our experience. Mindfulness helps us to stop spiralling in this way.

Another important fact to keep in mind is this: our internal experiences don't last forever. If, while you're taking a break from your chosen dopamine-driven activity or behaviour, you have the urge to re-download TikTok, or go on an online shopping spree, or get rip-roaringly drunk, tell yourself that the urge is *temporary*. No-one has ever had the experience of an urge lasting indefinitely. *It will pass.*

The urges and cravings are trying to tempt us towards old patterns, so it's important to remind ourselves that the uncomfortable experiences are fleeting. They are not permanent.

The same applies to those uncomfortable emotions that we wish to avoid or escape. If we feel bored or angry or sad, the desire to push away these feelings will mean we're more susceptible to picking up the video game controller or opening a bottle of wine. But it doesn't really work. The renowned Swiss psychiatrist, Carl Jung, said that whatever we resist 'persists'.[9] In other words, the more we push away these emotions, the more likely they are to come back. Recognising the temporary nature of urges and unpleasant emotions is paramount for cultivating self-awareness. It helps us feel more in control.

The reason this chapter is called 'Riding the wave' is because urges and emotions are essentially like waves. They start small, build and get stronger over time, reach a peak, but eventually they recede. Our job, using our mindfulness practices, is to *ride the wave*. If you're a visual person, imagine yourself on a surfboard, riding the wave of that internal experience. The surfer doesn't struggle with the wave. They don't try to hang onto it, nor do they try to struggle or fight it. They move *with* the wave, riding its natural rhythm. This concept applies to all urges and emotions, the pleasant ones as well as the unpleasant ones. Pleasant emotions will eventually fade into neutrality, while uncomfortable emotions will eventually dissolve back to their baseline.

Mindfulness is a skill, but it's also a way of being. It will help you as you go through your period of abstaining from your chosen dopamine-driven activity or behaviour. You'll experience feelings that are deeply uncomfortable but instead of reaching for a cigarette, a glass of wine or your phone, you'll sit with those feelings. You'll observe them curiously. You'll wonder whether they're just drawing you towards the target behaviour or something else entirely. Embracing the difficult or uncomfortable moments in this way allows us to tolerate them. It takes some of the power out of them and reduces suffering. It allows for us to have greater choice and control over what we do, rather than functioning on autopilot.

Once you have practised these techniques and learnt how to notice unpleasant internal experiences and sit with them rather than getting drawn into them, it's time to move on to the next step: exchanging the time spent doing unhealthy dopamine-driven habits for meaningful, values-driven pursuits. Before we get to that, though, let's take a look at what you can do if you really can't sit with your unpleasant feelings and the pull towards dopamine-driven behaviour feels irresistible.

LET'S REFLECT

- How comfortable are you with embracing discomfort?

- Are there specific situations where you find it easier or harder to tolerate discomfort?

- Reflect on your capacity to observe and name your own emotions. Is it something that you can do easily, or will it take some practice?

- How do you typically react when faced with difficult emotions? Are you more inclined to avoid them, distract yourself or confront them directly?

- Think about a time you may have tried mindfulness or a similar practice. What were your expectations and how did they align with your actual experience?

Chapter 18

When you can't
ride the wave

Riding the wave is a great skill that can help us to get through the inevitable challenges of life. But no two people are the same and, unfortunately, some do endure more hardship, loss, struggle and despair than others. Life – to use a hackneyed but truthful cliché – ain't fair.

On top of that, we are all genetically and biologically different. Emotional sensitivity exists on a spectrum, and what might feel like a small wave to some, is a massive dumper to others. At times, it might feel like riding the wave is exhausting, and you just want to jump off the surfboard, swim back to shore and lie down in the sun. Fair enough! Being able to sit with urges and emotions is an effective way to handle discomfort, but the reality is that we won't always have the energy or mental capacity to do so.

Say for example you're in a meeting at work. It's getting towards the end of a long day, but you still have a few hours ahead of you. You're tired, a bit hungry and frustrated at how the meeting is going. Suddenly, you remember the football game that's on tonight. You feel a strong urge to check the odds and see how your friends might be betting. You start thinking that once the meeting is over, you could pull out your phone and place a bet. However, sports betting is the behaviour you've chosen to take a break from. Checking the odds and chatting to friends who gamble is a risky thing to do, and you have a few hours of work left ahead of you that you need to focus on.

In this scenario, using mindfulness skills and noticing the dopamine urges, observing the experience, and sitting with the discomfort, is a perfectly reasonable way to try to allow this urge to pass by. However, it might not pass by that easily and you'll keep getting hit with wave after wave of dopamine firing, pulling you to re-engage with the behaviour you're meant to be abstaining from. Or maybe you feel so exhausted that you don't have the energy to ride those waves at all. Maybe being in that meeting, you need to put all your efforts into quickly and effectively managing that urge so that you can focus. You find yourself caught in an internal battle between dopamine-driven behaviours and value-led pursuits.

A client of mine, 46-year-old Oliver, was always dressed well and had a full head of hair. He was both charismatic and mysterious – the sort of character who really drew you in. Oliver came to see me because he was addicted to sex. It constantly occupied his thoughts and he'd go out on the weekends and spend a lot of money trying to impress women who might want to have casual sex with him. He also frequently attended brothels and hired escorts. Add drinking and drugs into the mix, and his problem only worsened.

Despite the sex, Oliver was lonely. The truth is, he wanted a loving, healthy relationship. He was divorced and missed having a partner with whom he could share his life, have a laugh over dinner, and hold close during the night. He yearned for companionship that went beyond physical intimacy. Oliver's problem was twofold. Firstly, he wasn't acting in ways that aligned with his value of having a loving, healthy relationship. He was pursuing hedonic happiness, rather than eudaimonic satisfaction. Secondly, when he did go on dates with women, with the view to getting to know them and forming a potential relationship, he'd always be drawn towards the pursuit of sexual gratification. This meant that at the end of the date, he'd try to convince the woman to have sex. If that didn't work, he'd visit a late-night massage parlour on his way home. He felt a lot of shame and regret about his behaviour and wanted to stop, but the dopamine pull was too strong. Oliver was fighting against a six-lane highway in his brain. He knew all the local brothels and erotic massage parlours and he knew how to hire an escort. He'd used them many times before, so this was a quick and easy thing for him to do. Resisting the urge felt infinitely harder.

'Riding the wave' wasn't enough for Oliver, especially when he was in the middle of a date with wandering thoughts. He experienced a clash between his values and his dopamine drive. He needed strategies that were quick and easy to enact in the heat of the moment, to keep him connected with his overarching goals and values.

So, what can we do when riding the wave isn't enough?

Luckily, we psychologists have a toolbox full of strategies that can help us regulate our emotions and manage discomfort. Some might be familiar, while others might be new. In this chapter, you'll learn some short, sharp and snappy exercises that you can use in all different scenarios and settings to help you cope when that dopamine-driven behaviour rears its ugly head.

Distraction

I'm sure I won't be telling you anything new by suggesting that distraction is a way to get through difficult experiences. Distraction can be an exceptionally helpful short-term tool when used in moderation. If we repeatedly use distraction when faced with something challenging, it can turn into an unhealthy pattern of avoidance, and we certainly don't want that! If we excessively distract ourselves we don't build the skill of being able to ride *some* waves. Suddenly every emotion or urge will feel so uncomfortable that we'll become reliant on distracting ourselves to avoid it. In fact, avoidance of pain can be one of the key factors that drives the development of an addiction, or a problematic relationship with a behaviour. Drinking, drugs, sex, porn, shopping, gambling – they're all ways that we can ultimately avoid a part of our life, and a part of ourselves.

Now, before we look at different specific ways to distract ourselves, I want you to first come up with some ideas yourself. All of us will at some point use a form of distraction to help us get through a difficult time. Take a minute to brainstorm what those methods of distraction are for you. Once you've come up with a few of your own ideas, here are some mental distractions and games to add to your list.

Mental games

- Go through the alphabet and for each letter come up with the name of an animal. A is for anteater, B is for bulldog, C is for caterpillar . . . and so on. If you don't like animals or have already done that, then you can try the same letter exercise with girls' names, boys' names, countries or cities around the world, colours, instruments, vegetables, sports, celestial objects . . . the sky is the limit!

- Mental arithmetic can be challenging and not always fun (if you ask me!), but it can certainly distract from the distress and help draw attention to the numbers. Try counting backwards from 100 by multiples of seven, or making up an addition chain. Start with a random number like five and then add a specific number to it each time such as three. Continue with the sequence mentally (five (plus three), eight (plus three), eleven, fourteen . . . and so on).

- Word associations can also help us ground ourselves by taking in the sights around us, while also playing a mental game. Look around you and choose an object. Now find three words that you can associate with the original word. For example, in front of me I have a pile of 'envelopes'. Three words that come to mind are 'mail', 'doctors' (because in the envelopes are letters to our clients' doctors) and 'stamps'. I also just recently got back from a trip to the Northern Territory, and I can see the didgeridoo that I bought in one corner of the room. Three words that come to mind are 'sound', 'colours' and 'land'.

- You need a page with some words on it handy for this next game. A book, a magazine, a newspaper, a flyer, a business card, even an old receipt shoved away in your wallet will work. Anything will do (although the more writing the better as the exercise will last for longer). I want you to look at that page and count all the letter e's. Every single one of them. Don't read the text, just count the e's. Not a fan of e's? Then how about all the p's, or any letter of your choosing!

Distraction is one of those tools that we can use quickly and easily. Trying to brainstorm ideas when we are in the middle of experiencing an urge or craving for a cigarette, or wanting to jump on social media, isn't the time to be considering loads of options available to us. When we're overwhelmed, our brain just doesn't think as rationally. Therefore, I highly recommend you have a list somewhere handy of a few simple strategies with which you can distract yourself when you're in the throes of discomfort and simply can't 'ride the wave'.

Use the senses

When our brain is activated by an urge or an uncomfortable inner experience, sensation or emotion, it can sometimes be hard to 'think' our way through. Our brain has less capacity to think rationally and we enter a more emotional state. Our level of arousal (or alertness) can also get in the way of us thinking clearly. Remember, when confronted with a problem, we don't have to solve it immediately. Instead, we can utilise 'sensory' strategies to manage our inner experience, lower our level of arousal, and then think our way through the issue using problem-solving strategies. This gives us the best chance of success to manage a difficult situation effectively.

If you're taking a break from your chosen dopamine-driven behaviour and notice strong urges or cravings to do it again, try some sensory strategies first. There are two general ways we can make use of the senses, and based on my clinical experience, people tend to have a preference for one or the other. We can go with very gentle, calming sensory experiences that help us re-regulate by soothing, or we can choose quite intensive strategies that provide more of a shock to our system. Let me outline them in more detail.

Self-soothing

These strategies are calming because they trigger the 'parasympathetic' nervous system which counteracts the body's stress response. It results in a decreased heart rate, reduced blood pressure and less muscle tension. In doing this, the body will gradually return to a state of equilibrium. Some soothing sensory activities can also stimulate the release of endorphins, which act as natural painkillers and mood enhancers, further promoting the feeling of relaxation.[1]

Think of the five senses – sight, sound, touch, smell and taste. For each of these, we'll consider various activities you can do that will provide a soothing and calming experience. Here are some ideas.

- **Sight:** Dimming lighting or putting on sunglasses can help calm us. Or try looking at beautiful images in magazines or coffee table books of vast landscapes, calming nature scenes and cities from around the world. Look at pictures of beautiful artworks. Watch videos on YouTube of natural scenery, or even close your eyes and visualise yourself in a peaceful and serene setting, like a tranquil beach, imagining the sights, sounds and sensations. Colours can also be soothing and doing some colouring or drawing using mindfulness colouring-in books can help with re-regulation. Alternatively, if you're up for an arts and craft day, make yourself a glitter jar (these can also be bought if you're not up for crafts!).

- **Sound:** Gentle classical music can be particularly calming (I love Mozart who is lighter than some other composers). Ambient music is also good, as it emphasises tone and atmosphere without the traditional beats, rhythms or structured melodies in other music. The layers of sound and texture can help with both active and passive listening and promote a sense of calm. Search nature sounds on YouTube and you'll find five-hour-long playlists of gentle rain, waves lapping on the shore, rustling leaves or chirping birds. If nature sounds aren't your thing, try white noise, pink noise or brown noise. White noise is essentially sound that contains all frequencies audible to the human ear, at equal intensity. It almost functions like an audio 'blanket' by masking disruptive sounds in the environment. Pink and brown noises are similar but without the higher frequencies. And if none of the above tickle your fancy, try recordings of autonomous sensory meridian response (ASMR), binaural beats, Tibetan singing bowls or monks chanting.

- **Touch:** When considering touch, think about texture, pressure and temperature. Warmth from a heat pack or a warm bath can be particularly soothing. Even being under the doona or wearing a big jumper can provide the comfort of physical touch as well as warmth. Many people find weighted blankets to be helpful, as they provide gentle pressure across the body, activating the parasympathetic nervous system. Massaging parts of your body can assist with touch as well as pressure. Some people love the feeling of different soft fabrics against their skin, like silk or velvet. You could also try stress balls, fidget spinners and cubes, and a particular favourite for kids – slime!

- **Smell:** Aromatherapy is essentially the practice of using aromatic plant oils to create scents which bring about physiological and emotional changes. Lavender oils, or incense with chamomile or sandalwood, help us to feel calm and re-regulate the nervous system. If you don't have a diffuser, inhaling a few drops on a cotton ball will do the trick. Alternatively, scented candles are wonderful, and nowadays, there are so many beautiful varieties to choose from (in fact, I bought a delicious one at Christmas that smelt of gingerbread!). You could also try scented lotions, herbal pillows, room sprays, scented bath salts, brewing a pot of herbal tea, or even scented gel pens.
- **Taste:** Nostalgic flavours can remind us of particular moments or times in our life and elicit feelings of calm and relaxation. Warm drinks like cups of tea, warm milk or even soups and broths all work too. Perhaps it's the taste of a small piece of chocolate (even better if it's dark chocolate with a high cocoa content, as it contains antioxidants and compounds that stimulate the release of endorphins).

People engage those senses they find most calming and beneficial. For me, it's touch and, in particular, the feeling of heat. When I'm having a bad day, I find putting a jumper on to be very comforting (even in summer!). A warm to hot shower can also provide me with relief and the ability to re-regulate my emotions. On the other hand, I dislike things that smell (aside from the smell of a delicious meal, that is), so perfumes, essential oils and incense are never going to calm me. If anything, they make me feel even more uncomfortable!

You can also mix and match your sensory experiences. Brewing a cup of tea might tick off smell, taste and temperature all at once. You could close your eyes and visualise a relaxing scene in nature (sight) while also listening to a recording of birds chirping (sound). If a warm bath is a way to relax (touch and temperature), then consider adding some scented bath salts (smell) and put on some ambient music in the background (sound).

Try to have sensory tools available that are quick and easy to use, no matter where you are. Preparing a playlist with calming sounds and music that you can listen to while out and about is great. The same goes for carrying a small bottle of essential oil, a fidget spinner or a piece of dark chocolate.

Intense sensations

We've talked about using the senses to soothe ourselves. But what about intense sensory experiences that also reset the nervous system? There are a few of those available to us as well.

Have you ever had a cold shower? Or an ice bath? What did it feel like? I imagine the average answer isn't 'pleasant' or 'relaxing', but in fact, ice-cold water can be extremely beneficial for managing cravings, urges or strong and distressing emotional experiences. The shock to the system is great for our body. Pretend you're intensely craving some chocolate, an ice cream, or a hazelnut-cream-filled doughnut. Or perhaps you're full of anxiety or anger at a situation. Imagine now stepping into a cold shower or jumping into a swimming pool. How do you think that would feel? It won't make you happy, but it'll serve as a circuit breaker to whatever you were experiencing a moment before!

Wim Hof, also known as 'The Iceman', is an extreme athlete from the Netherlands, known for his ability to withstand cold temperatures.[2] He previously held a Guinness World Record for swimming under ice, as well as a record for a barefoot half marathon on ice and snow (disclaimer – please *don't* try that at home!).

Thankfully, for the rest of us, we don't need to shoot for a Guinness World Record. We can achieve similar results just by taking a cold shower.

Ultimately, we are trying to mimic what's known as the 'deep dive reflex', or the 'mammalian dive reflex'. This is a physiological response that occurs in mammals, including humans, when we're submerged in cold water. When our body comes into contact with cold water, several physiological changes occur. Our heart rate slows which helps conserve oxygen and redirect blood flow, the blood vessels in our limbs constrict, and we slow down our breathing, or hold our breath if we're underwater.[3] If you can't face a cold shower, fill a bowl or basin with cold water and submerge your face into it, with ice cubes for added effect!

As I said earlier, most people tend to go either for the intense sensations or the gentle self-soothing. If the thought of a cold shower excites you, then this is for you. If you can't think of anything worse, then try a warm bath first! Other strong, sensory experiences which can provide a shock and help break us out of an urge or craving can be eating something with very strong flavours such as wasabi or hot sauce. Little packets of wasabi are easy to carry around in your wallet in case you want to use this technique when out and about. Ever chewed on a raw piece of ginger? It's not pleasant but it sure is intense! For some people, strong peppermint flavours or scents do the trick. As do smelling eucalyptus oil, sucking on sour lollies like warheads, eating popping candy or even a lemon. Intense exercise can also work – a vigorous run or some rapid star jumps on the spot!

When I worked with Oliver, my client who was struggling with a sex addiction, he was at first sceptical about how useful these strategies would be. However, he gave them a go and, to his surprise, he found them to be effective. Oliver preferred intense sensations. If he was out on a date and felt the urge to pursue old habits, he would excuse himself and ask the bar staff for a glass with some ice cubes in it. He would then go to the bathroom and use this to help regulate himself, holding the ice in his hands and rubbing the cubes on his face. If he was driving home after a date and started experiencing strong urges to take a detour to a brothel, he'd play some really loud, intense metal music (think Black Sabbath and Iron Maiden). He'd also carry little packets of wasabi in his wallet as an added intense sensory experience if needed.

All of the above suggestions are designed to function as a circuit breaker during a moment of distress. They won't fix a problem, nor are they long-term solutions. They're for when sitting with the urge or the distress — riding the wave — feels too overwhelming.

Consequential thinking

When you feel that dopamine-driven urge to enact the behaviour you're taking a break from, thinking about the potential negative outcomes can be a good and useful deterrent. It helps us anticipate the potential consequences of acting on that urge based on our past experience.

This strategy works best when we do some preparation. I always recommend that people write out a list of consequences in advance, because just as with sensations, it can be hard to think through things clearly when we're struggling with an urge. Keeping this list somewhere handy is ideal for when we are out and about. After you've brainstormed some potential consequences, consider the questions below to help you identify any additional long-term and short-term consequences.

- What's been the *worst* outcome from engaging in your dopamine-driven behaviour?
- What's been a *common* negative outcome from engaging in that behaviour?
- What could be a *potential* negative outcome?

- What are the *short-term* negative consequences?
- What about the *longer-term* consequences?
- Has this behaviour affected your health?
- Has it affected your mental health, wellbeing, or mood?
- Has it resulted in strained relationships?
- Has it resulted in you missing out on any opportunities or experiences?
- Has it affected your work or ability to focus?
- Has it affected your self-esteem or self-worth?
- How will you feel about yourself if you give into the urge?

The consequences we end up listing are not just practical, but also emotional.

I did this exercise with Oliver and he used the list when out on a date, or driving past a familiar brothel. His list included:

- Feeling guilt and shame about going back to old habits and behaviours that I want to abstain from.
- Feeding the preoccupation I have with sex. The more I chase it, the more I think about it.
- I end up feeling more alone and isolated, and always hate the feeling of going home alone afterwards.
- I end up having late nights, which means I'm tired for work the next day and not as productive.
- I also end up ruminating about the night before which means I don't focus as much at work.
- I've developed a reputation among my friends that I am actively trying to change.
- I will feel worse about myself if I act on the urge.
- The financial cost adds up.

You can either end the exercise there, or also consider the flip side. What are the *positive* consequences of resisting the urge and continuing the break from your chosen behaviour? Again, consider the questions below to help you come up with a list.

- What have been the positive outcomes so far from taking a break from your chosen dopamine-driven behaviour?
- How have you felt about yourself in resisting engaging with that behaviour?
- What positive effects have you noticed on your health?
- Have there been any positive outcomes in terms of your focus? Productivity? Energy levels? Relationships?
- What experiences have you been able to embrace or enjoy as a result of taking a break?
- Are you able to more meaningfully engage with your values by resisting the urge?
- How has your overall quality of life been affected?

For Oliver, there were many positive consequences of abstaining. These included feeling like he had more control over himself and his choices, and feeling more aligned with his pursuit of a meaningful, long-term relationship. He also said that abstaining would help him feel a greater sense of self-esteem, get better sleep, and have more mental space for other things in life, including other healthy relationships.

Reflecting on the negative consequences of going back to the behaviour, and the positive consequences of abstaining, helps us 'ride the wave' of the urge more effectively. Having these useful, realistic reminders helps us find that extra bit of inner strength to allow the urge to take its natural course, rising up in intensity, and eventually receding.

Hopefully this chapter has provided you with some additional tools to help you get through those challenging moments. Taking a break from something that we've done repeatedly, and for a long time, is *not* easy! It's really hard just to 'stop' something. Sitting with the discomfort and being mindful are crucial. It helps us build our overall capacity to tolerate discomfort without avoiding or escaping it. However, I'm an advocate for having multiple tools in the toolbox, and distraction, sensory experiences and consequential thinking can be very useful when you need some extra help.

LET'S REFLECT

- Consider where you might lie on a spectrum of emotional sensitivity. Are you someone who feels things often and deeply, or does it take something significant to happen to you in order to feel more emotional?

- What methods of distraction have you successfully used?

- Consider your own sensory preferences for self-soothing. Which senses do you find most calming and beneficial?

- Reflect on a recent urge or craving you experienced. How might applying the techniques discussed in this chapter have helped you manage that urge more effectively?

Chapter 19

Building the new

As you abstain from your target behaviour, you might find that some of your energy is naturally spent resisting the urge to throw this book in the bin, re-download TikTok, drive through a McDonald's or buy a six-pack of beer. This is where we must use our strategies from the previous chapters – sitting with discomfort, riding the wave, distraction, self-soothing and considering the consequences. Doing this takes willpower but that in itself is not enough. When something is taken away from someone, it's important to replace it with something else. Otherwise, the change is likely unsustainable and a person may very quickly revert to their old habits.

Sometimes, those dopamine-driven behaviours can serve as a bit of a 'lifejacket' for us. They can help us get through tough times. However, it becomes problematic when they do more harm than good. If we just take away a person's lifejacket and throw them in the deep end, things won't go well.

In order to live a life that is meaningful and fulfilling, we need to replace those dopamine-driven behaviours with values-guided actions and goals. This is where, as Socrates says, we focus on 'building the new'. When we feel connected to our values and goals, we're more motivated to confront obstacles and navigate life's ups and downs. Therefore, the more we invest in engaging in those values-aligned behaviours we set out for ourselves in Chapter 13, the less powerful the pull will be to return to the dopamine-driven urges.

We are going to use these goals in both 'proactive' and 'reactive' ways. To do this effectively, we need two different types of goals. Firstly, we need goals that are structured and planned ahead of time. For example, the Sunday night roast dinner with your parents and siblings that aligns with the value of 'family'. However, the Sunday roast dinner won't help me when I come home from work after a hard day and am about to have a Bridget Jones moment with a tub of ice cream. It's at that moment that I need a 'place filler'. These actions will still be values aligned, but they'll be *flexible* too. For example, instead of being pulled towards the tub of ice cream, I'll choose another behaviour which is aligned with my value of family, such as calling my mum for a chat.

If your value is creativity or learning a new skill, a larger goal might be to take guitar lessons every week on a Thursday evening, while the 'place filler' that is easier and quicker to access might be to practise the guitar at home, watch some YouTube videos of guitar technique, or make a Spotify playlist with new songs you'd like to learn.

If your value is religion or spirituality, the larger goal might be to attend church every Sunday and go to Bible study, while the 'place filler' might be to say a prayer, read a passage of the Bible, or listen to hymns.

If your value is health and fitness, attending bootcamp twice a week is a larger practical goal you can set for yourself, while a 'place filler' might be to do twenty squats in the living room.

It could be argued that the place fillers are essentially a form of distraction. This is true. What's the difference between listing all the boys' names you can think of from A through Z like I suggested in Chapter 18 and calling your mum or playing guitar? The place fillers allow us to achieve two things in one go. Yes, they can distract us from uncomfortable experiences, but they also take us closer to our meaningful goals and ultimate fulfilment. They remind us that when emotions and urges feel too overwhelming to sit with, there are meaningful actions we can take to help us feel grounded in our values, our identity and our sense of self. We don't have to avoid or escape the pain through other means.

Make a list of your 'place filler' goals, based on your values from Chapter 13, and keep them close at hand, perhaps in your phone or on a piece of paper in your wallet. When you feel the pull towards the behaviour that you're taking a break from, grab your list. Sometimes, it might take more than one place filler behaviour to help us deal with the urge. We might need to call our sister (value = connection), complete another level learning Spanish on Duolingo (value = achievement), then play some music and dance around the house (value = fun and spontaneity).

In addition to actually practising the behaviours discussed above, take some time at the end of every day to reflect on and consolidate the progress you have made. Ask yourself these questions:

- What am I proud of myself for today? Big or small, it all counts.
- What did I do today that was aligned with my values? Which choices aligned with which values?
- What can I consciously choose to do tomorrow which will also align with my values?
- Was there anything that got in the way of me living in alignment with my values today, and if so how can I overcome this next time?

Reflecting on these questions each day will help keep your goals front of mind and allow you to reset and regain focus. Hopefully there will be many days where you will reflect on these questions and feel pleased with what you've achieved.

Of course, that won't happen every day. There will, without a doubt, be days where you are busy, tired, stressed, lonely, forgetful or lazy. This is all part of the human experience. Change and progress are not linear, so expect some hiccups along the way. When you have a day where you've strayed from your values, treat yourself with compassion rather than harshness.

Self-compassion is a fundamental part of how we learn to relate to ourselves. It is, however, a practice with which many people struggle. It involves treating ourselves with kindness, understanding and acceptance, especially during times of hardship and failure. Self-compassion is a way of relating to ourselves without harsh judgement. In the same way that we'd feel compassion for someone else who's struggling, we must learn to have that same compassion for ourselves.

Professor Kristin Neff from the University of Texas is a trail-blazer in the world of self-compassion.[1] She explains it as comprising three key components:

1. **Self-kindness:** Being kind and understanding oneself in times of pain or failure, rather than being overly harsh and critical.
2. **Common humanity:** Understanding that our experience is part of the larger human experience of joy and suffering, rather than seeing it as separate or isolating.
3. **Mindfulness:** Holding painful thoughts and feelings in balance, rather than attaching or over-identifying with them.

When we are overly self-critical, we risk losing focus on our values and goals. In fact, strong self-criticism is a risk factor for us relapsing back into old habits and engaging with dopamine-driven urges.

Criticising ourselves never feels good. It creates an internal experience that we naturally want to avoid or escape, and what better way to do that than indulging in a dopamine-driven behaviour that enables us to feel some temporary relief! That's not to say that we should never reflect on the ways we could improve ourselves or on mistakes we have made. Approaching such reflection with harshness rather than understanding and encouragement is detrimental to our mental health and our overall goals.

Just as we explored where our values come from in Part 2, it is important to consider where the compassion or harshness we hold towards ourselves comes from. Self-compassion and self-criticism sit on a spectrum. Often the way we relate to ourselves is internalised from a young age. We are influenced by our parents, families, teachers, peers, culture and society. We may have grown up in a family that was highly critical not just of us, but everyone else. Perhaps we heard our mother regularly criticise herself if she gained weight. Perhaps we had teachers with very high expectations who meted out harsh punishments if we did not achieve them. Perhaps we grew up in a culture that emphasises self-criticism. Or perhaps we internalised messages from society such as 'no pain no gain', 'sleep when you're dead', 'boys don't cry' or 'if you're not busy, you're not trying hard enough'.

Research shows that self-compassion is good for us. It allows us to connect with feelings of common humanity, thereby allowing us to feel less alone and isolated in our struggles.[2] It's been linked to positive mental health and decreased psychopathology in both adults and adolescents.[3] It's also been shown to help with resilience and strength when we're faced with a range of life stressors, including divorce,[4] life transition, parenting children with illnesses and disabilities,[5] chronic health issues[6] and bullying.[7]

Let's turn our attention back to Socrates' timeless wisdom. 'The secret of change is to focus all of your energy not on fighting the old, but on building the new.' I hope by now you have a solid framework for how to 'build the new' including taking some time away from your chosen behaviours, riding the wave of discomfort using mindfulness skills, utilising distraction, sensory experiences and consequential thinking when the urges feel a bit too much, and choosing meaningful pursuits to replace the dopamine-driven habits. Use those mindfulness skills to continue to *notice* how it feels to choose these pursuits over other, short-term pleasurable ones. Each meaningful distraction, engagement with values-aligned actions and act of self-compassion brings us closer to living the life we desire for ourselves.

So now what? What do you do next? Read on to learn how to make decisions about if and how you might reintegrate your target behaviour into your life.

LET'S REFLECT

- Can you think of a time where you were able to implement a new habit or make a long-lasting change? What strategies did you use to make this change sustainable?

- How do you currently incorporate self-reflection into your daily routine? Are there any specific questions or prompts that help you stay focused on your values and goals?

- Do you tend to approach yourself with kindness and understanding, or do you often default to self-criticism?

- How can you cultivate a greater sense of self-compassion?

Chapter 20

Breaking free and finding balance

You've made it to the final chapter – congratulations! Whether you've been through the process and implemented the strategies or read the book considering whether you might like to make some changes, well done. Simply by getting to the end, you've made a profound change to your life, to your knowledge and to your options for yourself.

From here on, the choice is yours. Once you've finished your break from your chosen substance or behaviour, you can now consider what you want to do next. Do you want to continue with the break? Do you want to slowly reintroduce the substance or the behaviour? Or do you want to go back to doing exactly what you did before? It's up to you. However, you hopefully get to make this choice *consciously*, with some new insights as to how this specific substance or behaviour influences your life.

Consider what new perspectives you've gained by taking a break from your dopamine-driven behaviour. What have you learnt about how you live life and the choices you make? How many of your day-to-day decisions are made while on autopilot, and have any of these become more conscious? Consider what it's felt like to be able to implement more meaningful, values-based actions into your week.

It's a cliché, but life is tough! From navigating professional and personal challenges, to grappling with societal pressures, life can sometimes feel like an endless, uphill battle. It's only natural to yearn for moments of respite and a temporary space away from the demands of reality. Seeking these moments isn't just a desire, it's a common human experience. We all need time to recharge, reset and rejuvenate, allowing us to gain perspective and face the highs and lows. It's like hitting the pause button every so often.

The challenge lies in striking a balance. Making change is hard. It requires patience, tolerance, the ability to sit with uncertainty, and self-forgiveness. But avoiding reality doesn't work and it simply isn't possible. Eventually, whatever we are trying to avoid will resurface, often with greater intensity. The more we can face the challenges in life, the better we become at navigating them and sitting with the discomfort they inevitably bring. We learn to accept that happiness is a fleeting goal, and that we will never be happy all of the time. But we can instead learn to ride the waves of life, without getting too attached to the highs and lows.

You may be wondering what happened to my clients and how their situations turned out.

Oliver (from Chapter 18), who was engaging in compulsive sexual behaviours, saw me for a few years. It was a long process for him and there were periods where he'd cancel sessions, disappear and then re-emerge months later. The dopamine-driven urges would become strong again and he'd stop therapy. But he always came back. He hit a few stumbling blocks along the way, but he managed to learn and practise the skills needed to resist the urges to go back to casual sex and escorts. Oliver met a wonderful lady with whom he developed a solid relationship. I met her on two occasions, when he invited her into the therapy room so she could get a better understanding of what he had been grappling with. I still see Oliver from time to time for the purposes of maintenance and relapse prevention. For the most part, he is healthy, content in his relationship and happy in himself.

Ben (Chapters 8 and 11) didn't suffer from an addiction as severe as Oliver's. He made some adjustments to his life, planning some travel with friends, joining the university basketball team and forming a new band. He decided to continue with accounting and eventually went back to long-distance running after taking a break, but this time with more balance in his life. When Ben and I parted ways, he'd booked a trip to see his best friend in the US. I'm assuming that went ahead. I finished my sessions with Ben as I do with all my clients – with an open offer for him to return at any point in the future, should he need additional support. I haven't heard from him and in my line of work no news usually means good news!

Belinda, the burnt-out investment banker from Chapter 10, transitioned into a new, healthier role, realigned her values and had a baby.

Garry from Chapter 6 was able to find balance in his conflicts between pleasure and purpose. His high-flying lifestyle continued, though he learnt to moderate this with effectively integrated, meaningful pursuits around culture, connection and giving back to his community.

Sandra from Chapter 6 found balance in her life by stepping away from the need to provide for others all the time. She found fun and creative ways to experience short-term pleasure through art and comedy.

Joe from Chapter 15 who compulsively used pornography was able to follow the suggestions in this book and change his relationship with pornography while dedicating more meaningful time to his family and hobbies.

And Molly, my client who had a history of childhood trauma from Chapter 17, was able to utilise mindfulness and other cognitive and emotional tools to help her work through and process her trauma. She broke free from the repetitive feedback loop that kept her stuck in those traumatic memories and was able to embrace her life in the here and now.

•

I truly hope reading *The Dopamine Brain* has given you a basic understanding of how this important neurotransmitter works. I hope you've found space to reflect on your values and acquired some skills to live a more fulfilling life successfully while resisting those dopamine pulls. As clichéd as it may sound, making positive change is a lifelong journey. It's okay to stumble along the way. Make sure to celebrate the 'small wins' and improvements and approach the inevitable setbacks with compassion and understanding. Keep your values close and let them guide you down a path towards a life that is rich in meaning and fulfilment. Yes, you'll need to revisit your values-based goals every so often and tweak them so that they fit in with your life. It's a fluid process. The aim is to make conscious choices about when and how you get pulled by those dopamine-driven urges, mindfully choosing behaviours and substances in moderation, rather than allowing them to control you.

As we conclude this journey, I encourage you to hold onto one final takeaway. *The brain is an incredible organ and it can change.* Use the knowledge you have acquired in this book to change it in your favour.

LET'S REFLECT

- Reflect on the insights you've gained through reading this book. How have these influenced your understanding of yourself and your choices?

- What has it felt like to implement more meaningful, values-based actions into your week?

- How can you continue to maintain the progress you have made in both increasing your values-based actions and decreasing your dopamine-driven behaviours?

- What might get in the way of maintaining progress and how can you overcome it?

Endnotes

The Dopamine Brain is grounded in scientific research. Throughout the book, I refer to numerous studies. If you wish to explore these further, the references are listed below.

Introduction

1   Ebbinghaus, Hermann, *Psychology: An elementary text-book*, Arno Press, New York, 1973.
2   Frey, Emil F., 'The earliest medical texts' in *Clio Medica. Acta Academiae Internationalis Historiae Medicinae*, Vol. 20, Brill, Leiden, pp. 79–90.

Chapter 1

1   Dum, Rachel et al., 'Dopamine receptor expression and the pathogenesis of attention-deficit hyperactivity disorder: a scoping review of the literature', *Current Developmental Disorders Reports*, 9(4), 2022, pp. 127–136.
2   Delva, Nella C. and Stanwood, Gregg D., 'Dysregulation of brain dopamine systems in major depressive disorder', *Experimental Biology and Medicine*, 246(9), 2021, pp. 1084–1093.
3   Olivares-Hernández et al., 'Dopamine receptors and the kidney: an overview of health-and pharmacological-targeted implications', *Biomolecules*, 11(2), 2021, p. 254.

4   Previc, Fred H., *The Dopaminergic Mind in Human Evolution and History*, Cambridge University Press, New York, 2011, Chapter 1.

5   Baldo, Brian A. and Kelley, Ann E., 'Discrete neurochemical coding of distinguishable motivational processes: insights from nucleus accumbens control of feeding', *Psychopharmacology*, 191, 2007, pp. 439–459.

6   Hull, Elaine M.; Muschamp, John W. and Sato, Satoru, 'Dopamine and serotonin: influences on male sexual behavior', *Physiology & Behavior*, 83(2), 2004, pp. 291–307.

7   Latif, Saad et al., 'Dopamine in Parkinson's disease', *Clinica Chimica Acta*, 522, 2021, pp. 114–116.

8   Salamone, J. D. et al., 'Beyond the reward hypothesis: alternative functions of nucleus accumbens dopamine', *Current Opinion in Pharmacology*, 5(1), 2005, pp. 34–41.

Chapter 2

1   https://www.forbes.com/sites/brucelee/2021/05/18/covid-19-vaccine-magnet-challenge-videos-claim-magnets-stick-to-arms-after-vaccination/?sh=52151f233421

2   Gabis, Lidia V. et al., 'The myth of vaccination and autism spectrum', *European Journal of Paediatric Neurology*, 36, 2022, pp. 151–158.

3   https://www.abc.net.au/news/2008-02-19/goldfish-three-second-memory-myth-busted/1046710

4   https://www.abc.net.au/news/2023-03-26/digital-detoxing-dopamine-fasting/102086330

5   Abdelnour, Elie; Jansen, Madison O. and Gold, Jessica A., 'ADHD diagnostic trends: Increased recognition or overdiagnosis?', *Missouri Medicine*, 119(5), 2022, p. 467.

6   Dougherty, Darin D. et al., 'Dopamine transporter density in patients with attention deficit hyperactivity disorder', *The Lancet*, 354(9196), 1999, pp. 2132–2133.

7   Spencer, T. et al., 'A large, double-blind, randomized clinical trial of methylphenidate in the treatment of adults with attention-deficit/hyperactivity disorder', *Biological Psychiatry*, 57(5), 2005, pp. 456–463.

8   Bowman, Elizabeth et al., 'Not so smart? "Smart" drugs increase the level but decrease the quality of cognitive effort', *Science Advances*, 9(24), 2023, eadd4165.

9 https://www.abc.net.au/news/2023-03-26/digital-detoxing-dopamine-fasting/102086330

10 Jones, Alexis et al., 'Identifying effective intervention strategies to reduce children's screen time: a systematic review and meta-analysis', *International Journal of Behavioral Nutrition and Physical Activity*, 18, 2021, pp. 1–20.

11 Houghton Stephen et al., 'Virtually impossible limiting Australian children and adolescents daily screen based media use', *BMC Public Health*, 15(5), 2015, pp. 1–11.

12 AAP Council on Communications and Media, 'Media and Young Minds', *Pediatrics*, 138(5), 2016, pp. 1–8.

13 Zeman, Janice et al., 'Emotion regulation in children and adolescents', *Journal of Developmental & Behavioral Pediatrics*, 27(2), 2006, pp. 155–168.

Chapter 3

1 Griffiths, Mark, 'A "components" model of addiction within a biopsychosocial framework', *Journal of Substance Use*, 10(4), 2005, pp. 191–197.

2 Paakkari, Leena et al., 'Problematic social media use and health among adolescents', *International Journal of Environmental Research and Public Health*, 18(4), 2021, p. 1885.

3 Oviedo-Trespalacios, Oscar et al., 'Problematic use of mobile phones in Australia . . . is it getting worse?', *Frontiers in Psychiatry*, 10, 2019, 440510.

4 https://www.who.int/standards/classifications/frequently-asked-questions/gaming-disorder

5 Starcevic, Vladan et al., 'Problematic online behaviors and psychopathology in Australia', *Psychiatry Research*, 327, 2023, 115405.

Chapter 4

1 https://www.who.int/europe/news/item/04-01-2023-no-level-of-alcohol-consumption-is-safe-for-our-health
Anderson, Benjamin O. et al., 'Health and cancer risks associated with low levels of alcohol consumption,' *The Lancet Public Health*, 8(1), 2023, pp. e6–e7.

2 https://www.nhmrc.gov.au/health-advice/alcohol

3   https://www.abs.gov.au/statistics/health/health-conditions-and-risks/alcohol-consumption/latest-release#:~:text=Guideline%201%20recommends%20that%20'To,risk%20of%20harm%20from%20alcohol'.

4   https://monographs.iarc.who.int/list-of-classifications

5   https://www.cancer.nsw.gov.au/prevention-and-screening/prevent-ing-cancer/reduce-your-cancer-risk/drink-less-alcohol#:~:text=It%20does%20not%20matter%20the,cancer%20increases%20with%20every%20drink.&text=Watch%20our%20video%20about%20alcohol,of%20cancer%20it%20can%20cause

6   Piano, Mariann R., 'Alcohol's Effects on the Cardiovascular System', *Alcohol Research*, 38(2), 2017, pp. 219–241.

7   Valenzuela, Fernando C., 'Alcohol and Neurotransmitter Interactions', *Alcohol Health and Research World*, 21(2), 1997, pp. 144–148.

8   Di Chiara, Gaetano, 'Alcohol and dopamine', *Alcohol Health and Research World*, 1997, 21(2), pp. 108–114. PMID: 15704345; PMCID: PMC6826820.

9   https://nymag.com/intelligencer/2018/01/does-instagram-withhold-likes-to-get-users-to-open-app.html

10   www.aihw.gov.au/reports/australias-welfare/gambling

11   Turner, Nigel E.; Zangeneh, Masood and Littman-Sharp, Nina, 'The experience of gambling and its role in problem gambling', *International Gambling Studies*, 6(2), 2006, pp. 237–266.

12   Calabrò Rocco S. et al., 'Neuroanatomy and function of human sexual behavior: A neglected or unknown issue?', *Brain Behavior*, 2019, 9(12), e01389.

13   Potenza, Marc N., 'Non-substance addictive behaviors in the context of DSM-5', *Addictive Behaviors*, 2014, 39(1) pp. 1–2.

14   Dwulit, Aleksandra D. and Rzymski, Piotr, 'Prevalence, patterns and self-perceived effects of pornography consumption in polish university students: A cross-sectional study', *International Journal of Environmental Research and Public Health*, 16(10), 2019, p. 1861.

15   Ybarra, Michele L. et al., 'X-rated material and perpetration of sexually aggressive behavior among children and adolescents: Is there a link?', *Aggressive Behavior*, 37(1), 2011, pp. 1–18.

16   Hartston, Heidi, 'The case for compulsive shopping as an addiction', *Journal of Psychoactive Drugs*, 44(1), 2012, pp. 64–67.

17 Balakrishnan, Janarthanan and Griffiths, Mark D., 'Perceived addictiveness of smartphone games: A content analysis of game reviews by players', *International Journal of Mental Health and Addiction*, 17(4), 2019, pp. 922–934.

18 Tekinbaş, Katie S. and Zimmerman, Eric, *Rules of play: Game design fundamentals*, MIT Press, Cambridge MA, 2003.

19 Weinstein, Aviv and Lejoyeux, Michel, 'Internet addiction or excessive internet use', *The American Journal of Drug and Alcohol Abuse*, 36(5), 2010, pp. 277–283.

20 Weinstein, Aviv and Lejoyeux, Michel, 'Neurobiological mechanisms underlying internet gaming disorder', *Dialogues in Clinical Neuroscience*, 22(2), 2020, pp. 113–126.

21 Tian, Ming-Yuan et al., 'Brain structural alterations in internet gaming disorder: Focus on the mesocorticolimbic dopaminergic system', *Progress in Neuro-Psychopharmacology and Biological Psychiatry*, 2023, 110806.

22 https://lighthouse.mq.edu.au/article/december-2023/screen-addicted-kids-become-screen-addicted-adults

23 Fino, E., et al., 'Factor structure, reliability and criterion-related validity of the English version of the Problematic Series Watching Scale', *BJPsych Open*, 8(5), 2022, e160.

24 Balakrishnan, Janarthanan and Griffiths, Mark D., 'An exploratory study of "selfitis" and the development of the Selfitis Behavior Scale', *International Journal of Mental Health and Addiction*, 16(3), 2018, pp. 722–736.

25 Atroszko, Paweł A. et al., 'Study addiction – A new area of psychological study: Conceptualization, assessment, and preliminary empirical findings', *Journal of Behavioral Addictions*, 4(2), 2015, pp. 75–84.

26 https://lighthouse.mq.edu.au/article/december-2023/screen-addicted-kids-become-screen-addicted-adults

27 https://theconversation.com/dating-apps-are-accused-of-being-addictive-what-makes-us-keep-swiping-224068

Chapter 5

1   Ali, Ansam et al., 'Endorphin: function and mechanism of action', *Science Archives*, 2, 2021, pp. 9–13.

2   https://injuryfacts.nsc.org/home-and-community/safety-topics/drug overdoses/data-details/

3   Brands, Bruna; Marshman, Joan A. and Sproule, Beth, *Drugs & Drug Abuse: A Reference Text*, Addiction Research Foundation, University of Minnesota, 1998.

4   Johnson, Zachary V. et al., 'Oxytocin receptors modulate a social salience neural network in male prairie voles', *Hormones and Behavior*, 87, 2017, pp. 16–24.

5   Russell, John A.; Leng, Gareth and Douglas, Alison J., 'The magnocellular oxytocin system, the fount of maternity: adaptations in pregnancy', *Frontiers in Neuroendocrinology*, 24(1), 2003, pp. 27–61.

6   Kosfeld, Michael et al., 'Oxytocin increases trust in humans', *Nature*, 435(7042), 2005, pp. 673–676.

7   Domes, Gregor et al., 'Oxytocin improves "mind-reading" in humans', *Biological Psychiatry*, 61(6), 2007, pp. 731–733.

8   Zak, Paul J.; Stanton, Angela A. and Ahmadi, Sheila, 'Oxytocin increases generosity in humans', *PLOS ONE*, 2(11), 2007, e1128.

9   https://www.who.int/groups/commission-on-social-connection

10 Botha, Ferdi, 'Social connection and social support', in Wilkins et al., *The Household, Income and Labour Dynamics in Australia Survey: Selected Findings from Waves 1 to 20*, The Melbourne Institute, Melbourne, 2022.

11 Holt-Lunstad, Julianne, 'Loneliness and social isolation as risk factors for mortality: a meta-analytic review', *Perspectives on Psychological Science*, 10(2), 2015, pp. 227–237.

12 Meltzer, Howard et al., 'Feelings of loneliness among adults with mental disorder', *Social Psychiatry and Psychiatric Epidemiology*, 48, 2013, pp. 5–13.

13 Tsai, Tsung-Yu et al., 'The interaction of oxytocin and social support, loneliness, and cortisol level in major depression', *Clinical Psychopharmacology Neuroscience*, 17(4), 2019, pp. 487–494.

14 Gershon, Michael D. and Tack, Jan, 'The serotonin signaling system: from basic understanding to drug development for functional GI disorders', *Gastroenterology*, 132(1), 2007, pp. 397–414.

15 Coppen, Alec, 'The biochemistry of affective disorders', *The British Journal of Psychiatry*, 113(504), 1967, pp. 1237–1264.

16 Pilkington, Pamela D.; Reavley, Nicola J. and Jorm, Anthony F., 'The Australian public's beliefs about the causes of depression: Associated

factors and changes over 16 years', *Journal of Affective Disorders*, 150(2), 2013, pp. 356–362.

17 Read, John et al., 'A survey of UK general practitioners about depression, antidepressants and withdrawal: Implementing the 2019 Public Health England report', *Therapeutic Advances in Psychopharmacology*, 10, 2020, 2045125320950124.

18 Moncrieff, Joanna et al., 'The serotonin theory of depression: A systematic umbrella review of the evidence', *Molecular Psychiatry*, 28(8), 2023, pp. 3243–3256.

19 Ferreri, L. et al., 'Dopamine modulates the reward experiences elicited by music', *Proceedings of the National Academy of Sciences*, 116(9), 2019, pp. 3793–3798.

20 Ferreri, Laura et al., 'Dopamine modulates the reward experiences elicited by music', *Proceedings of the National Academy of Sciences*, 116(9), 2019, pp. 3793–3798.

21 Dunbar, R. I. M. et al., 'Performance of music elevates pain threshold and positive affect: Implications for the evolutionary function of music', *Evolutionary Psychology*, 10(4), 2012, 147470491201000403.

22 Fritz, Thomas H. et al., 'Musical agency reduces perceived exertion during strenuous physical performance', *Proceedings of the National Academy of Sciences*, 110(44), 2013, pp. 17784–17789.

Chapter 8

1 Witteman, Holly O. et al., 'Clarifying values: an updated and expanded systematic review and meta-analysis', *Medical Decision Making*, 41(7), 2021, pp. 801–820.

2 Medeiros, Christina et al., 'Decision aids available for parents making end-of-life or palliative care decisions for children: A scoping review', *Journal of Paediatrics and Child Health*, 56(5), 2020, pp. 692–703.

3 Peinado, Susana et al., 'Values clarification and parental decision making about newborn genomic sequencing', *Health Psychology*, 39(4), 2020, p. 335.

4 Delaney, Rebecca K. et al., 'Study protocol for a randomised clinical trial of a decision aid and values clarification method for parents of a fetus or neonate diagnosed with a life-threatening congenital heart defect', *BMJ Open*, 11(12), 2021, e055455.

5   Knafo, Ariel and Schwartz, Shalom H. 'Identity formation and parent-child value congruence in adolescence', *British Journal of Developmental Psychology*, 22(3), 2004, pp. 439–458.

6   Schuster, Carolin; Pinkowski, Lisa and Fischer, Daniel, 'Intra-Individual Value Change in Adulthood', *Zeitschrift für Psychologie*, 227(1), 2019, pp. 42–52.

7   Oppenheim-Weller, Shani; Roccas, Sonia and Kurman, Jenny, 'Subjective value fulfillment: A new way to study personal values and their consequences', *Journal of Research in Personality*, 76, 2018, pp. 38–49.

8   Cohen, Geoffrey L. and Sherman, David K., 'The psychology of change: Self-affirmation and social psychological intervention', *Annual Review of Psychology*, 65, 2014, pp. 333–371.

9   https://www.ted.com/talks/jan_stassen_why_values_matter_jan_2019

Chapter 10

1   Hitlin, Steven, 'Values as the core of personal identity: Drawing links between two theories of self', *Social Psychology Quarterly*, 2003, pp. 118–137.

2   Erikson, Erik H., *Identity, Youth and Crisis*, W. W. Norton & Company, New York, 1968.

Chapter 11

1   Ardelt, Monika and Grunwald, Sabine, 'The importance of self-reflection and awareness for human development in hard times', *Research in Human Development*, 15(3–4), 2018, pp. 187–199.

Chapter 12

1   Harris, Russ, *ACT Made Simple: An easy-to-read primer on acceptance and commitment therapy*, New Harbinger Publications, Oakland, 2019.

Chapter 14

1   Hebb, Donald O., *The Organisation of Behaviour*, John Wiley & Sons, New York, 1949.

2   O'Brien, Charles P., 'Neuroplasticity in addictive disorders', *Dialogues in Clinical Neuroscience*, 11(3), 2009, pp. 350–353.

3   O'Brien, Charles P., 'Neuroplasticity in addictive disorders', *Dialogues in Clinical Neuroscience*, 11(3), 2009, pp. 350–353.

4 Maguire, Eleanor A.; Woollett, Katherine and Spiers, Hugo J., 'London taxi drivers and bus drivers: a structural MRI and neuropsychological analysis', *Hippocampus*, 16(12), 2006, pp. 1091–1101.

5 Pekna, Marcela and Pekny, Milos, 'The neurobiology of brain injury', *Cerebrum: the Dana Forum on Brain Science*, Vol. 2012, 2012, Dana Foundation.

6 Doidge, Norman, *The Brain That Changes Itself: Stories of personal triumph from the frontiers of brain science*, Scribe Publications, Melbourne, 2010.

7 https://people.com/mora-leeb-16-had-half-her-brain-removed-as-a-baby-glass-half-full-life-exclusive-8407334

8 Kliemann, Dorit et al., 'Intrinsic functional connectivity of the brain in adults with a single cerebral hemisphere', *Cell Reports*, 29(8), 2019, pp. 2398–2407.

Chapter 15

1 Lally, Phillippa et al., 'How are habits formed: Modelling habit formation in the real world', *European Journal of Social Psychology*, 40(6), 2010, pp. 998–1009.

2 Lembke, Anna, *Dopamine Nation: Finding balance in the age of indulgence*, Headline, London, 2021.

3 Childress, Anna Rose et al., 'Limbic activation during cue-induced cocaine craving', *American Journal of Psychiatry*, 156(1), 1999, pp. 11–18.

4 Childress, Anna Rose et al., 'Prelude to passion: Limbic activation by "unseen" drug and sexual cues', *PLOS ONE*, 3(1), 2008, e1506.

5 Volkow, Nora D. et al., 'Cocaine cues and dopamine in dorsal striatum: mechanism of craving in cocaine addiction', *Journal of Neuroscience*, 26(24), 2006, pp. 6583–6588.

6 Muckle, Wendy et al., 'Managed alcohol as a harm reduction intervention for alcohol addiction in populations at high risk for substance abuse', *Cochrane Library*, 12 December 2012.

7 O'Brien, Charles P., 'Neuroplasticity in addictive disorders', *Dialogues in Clinical Neuroscience*, 11(3), 2009, pp. 350–353.

8 https://link.springer.com/referenceworkentry/10.1007/978-3-642-27772-6_7019-1

9 Gawin, Frank H. and Kleber, Herbert D., 'Abstinence symptomatology and psychiatric diagnosis in cocaine abusers: clinical observations', *Archives of General Psychiatry*, 43(2), 1986, pp. 107–113.

10 Neisewander, Janet L. et al., 'Fos protein expression and cocaine-seeking behavior in rats after exposure to a cocaine self-administration environment', *Journal of Neuroscience*, 20(2), 2000, pp. 798–805.

11 Grimm, Jeffrey W., 'Incubation of food craving in rats: A review', *Journal of the Experimental Analysis of Behavior*, 113(1), 2020, pp. 37–47.

12 Vafaie, Nilofar and Kober, Hedy, 'Association of Drug Cues and Craving With Drug Use and Relapse: A Systematic Review and Meta-analysis', *JAMA Psychiatry*, 79(7), 2022, pp. 641–650.

13 Kulkarni, Kaustubh R.; O'Brien, Madeline and Gu, Xiaosi, 'Longing to act: Bayesian inference as a framework for craving in behavioral addiction', *Addictive Behaviors*, 2023, 107752.

Chapter 16

1 Freud, Sigmund, 'Beyond the pleasure principle', *Psychoanalysis and History*, 17(2), 2015, pp. 151–204.

2 Eisenberger, Naomi I., 'The neural bases of social pain: Evidence for shared representations with physical pain', *Psychosomatic Medicine*, 74(2), 2012, p. 126.

3 Kross, Ethan et al., 'Social rejection shares somatosensory representations with physical pain', *Proceedings of the National Academy of Sciences*, 108(15), 2011, pp. 6270–6275.

4 Kross, Ethan et al., 'Neural dynamics of rejection sensitivity', *Journal of Cognitive Neuroscience*, 19(6), 2007, pp. 945–956.

5 Thompson, Ross; Meyer, Sara and Jochem, Rachel, 'Emotional regulation', *Encyclopedia of Infant and Early Childhood Development*, 2008, pp. 431–441.

6 McLaughlin, Katie A. et al., 'Emotion dysregulation and adolescent psychopathology: A prospective study', *Behaviour Research and Therapy*, 49(9), 2011, pp. 544–554.

7 Aldao, Amelia; Nolen-Hoeksema, Susan and Schweizer, Susanne, 'Emotion-regulation strategies across psychopathology: A meta-analytic review', *Clinical Psychology Review*, 30(2), 2010, pp. 217–237.

Chapter 17

1 https://www.miriamrosefoundation.org.au/dadirri/

2 Kabat-Zinn, Jon, *Wherever You Go, There You Are: Mindfulness meditation in everyday life*, Hachette, New York, 1994.

3   Lieberman, Matthew D. et al., 'Affect labeling disrupts amygdala activity in response to affective stimuli', *Psychological Science*, 18(5), 2007, pp. 421–428.

4   Marikar Bawa, Fathima L. et al., 'Does mindfulness improve outcomes in patients with chronic pain? Systematic review and meta-analysis', *British Journal of General Practice*, 65(635), 2015, e387–e400.

5   Hölzel, Britta K. et al., 'Mindfulness practice leads to increases in regional brain gray matter density', *Psychiatry Research: Neuroimaging*, 191(1), 2011, pp. 36–43.

6   Boccia, Maddalena; Piccardi, Laura and Guariglia, Paola, 'The meditative mind: A comprehensive meta-analysis of MRI studies', *BioMed Research International*, 2015, 419808.

7   Shires, Alice et al., 'The efficacy of mindfulness-based interventions in acute pain: A systematic review and meta-analysis', *Pain*, 161(8), 2020, pp. 1698–1707.

8   Paschali, Myrella et al., 'Mindfulness-based interventions for chronic low back pain: A systematic review and meta-analysis', *The Clinical Journal of Pain*, 40(2), 2023, pp. 105–113.

9   Jung, Carl G., *The Basic Writings of CG Jung: Revised Edition*, Vol. 20, Princeton University Press, Princeton, 1990.

Chapter 18

1   Tindle, Jacob and Tadi, Prasanna, *Neuroanatomy, Parasympathetic Nervous System*, StatPearls Publishing, Treasure Island, Florida, 2020.

2   https://www.guinnessworldrecords.com/records/hall-of-fame/wim-hof-the-iceman

3   Godek, Devon and Freeman, Andrew M., *Physiology, Diving Reflex*, StatPearls Publishing, Treasure Island, Florida, 2024.

Chapter 19

1   Neff, Kristin D., 'Self-compassion: An alternative conceptualization of a healthy attitude toward oneself', *Self and Identity*, 2(2), 2003, pp. 85–101.

2   Bluth, Karen and Neff, Kristin D., 'New frontiers in understanding the benefits of self-compassion', *Self and Identity*, 17(6), 2018, pp. 605–608.

3   Marsh, Imogen C.; Chan, Stella W. Y. and MacBeth, Angus, 'Self-compassion and Psychological Distress in Adolescents – a Meta-Analysis', *Mindfulness*, Vol. 9, 2018, pp. 1011–1027.

4   Sbarra, David A.; Smith, Hillary L. and Mehl, Matthias R., 'When Leaving Your Ex, Love Yourself: Observational Ratings of Self-Compassion Predict the Course of Emotional Recovery Following Marital Separation', *Psychological Science*, Vol. 23, Issue 3, 2012.

5   Wong, Celia C. Y. et al., 'Self-Compassion: a Potential Buffer Against Affiliate Stigma Experienced by Parents of Children with Autism Spectrum Disorders', *Mindfulness*, Vol. 7, 2016, pp. 1385–1395.

6   Brion, John M.; Leary, Mark R. and Drabkin, Anya S., 'Self-compassion and reactions to serious illness: The case of HIV', *Journey of Health Psychology*, Vol. 19, Issue 2, 2013.

7   Múzquiz, Juan; Pérez-García, Ana M. and Bermúdez, José, 'Relationship between Direct and Relational Bullying and Emotional Well-being among Adolescents: The role of Self-compassion', *Current Psychology*, Vol. 42, 2023, pp. 15874–15882.

Acknowledgements

I'd like to extend my heartfelt thanks to all the people in both my personal and professional life who have supported me not only throughout the writing of this book, but throughout my journey so far.

The biggest thank you to my mum and dad, who provide me with unconditional love and unwavering support in all my adventures and wild ideas. I couldn't have possibly done this (or many other things) without you.

This book wouldn't exist without Ashwin Khurana. Ashwin initially emailed me in mid-2023 to ask if I'd be interested in writing a book with Penguin Random House Australia. I was surprised and somewhat sceptical to receive this email out of the blue from an editor of a major publishing house. After a quick google to fact-check that Ashwin was in fact a real person, my scepticism changed to elation. All I can say is that I feel so incredibly grateful to have received that email and have been offered this wonderful opportunity.

I'd like to extend my gratitude to Rod Morrison, whose expert editorial guidance has been invaluable. Thank you for your dedication and assistance. You helped me bring this book to life.

To the rest of the team at Penguin Random House Australia – Lily Crozier, Madison Du and Izzy Yates – I thank you for your encouragement and enthusiasm in making this book a reality.

I have been incredibly lucky to have had some of the most amazing mentors and supervisors throughout my career, who always encouraged me to pursue my passions and goals. A special mention to Professor Ian Kneebone (University of Technology Sydney), my PhD supervisor and an ongoing mentor, as well as Professor Rachel Roberts (University of Adelaide), Professor Toby Newton-John (University of Technology Sydney), Professor Frans Verstraten (University of Sydney) and Professor Ahmed Moustafa (Bond University).

To my friends and colleagues at the Australian Institute for Human Wellness, thank you for the encouragement along the way and for celebrating the milestones of writing this book with me.

I'm especially grateful for the time I spent working at St John of God Hospital, a psychiatric hospital in Sydney, where I ran the outpatient addictions program. This job, combined with my research, made me realise my passion for helping those who were struggling with addictions, as well as build awareness in the community about risks and harms of addictive substances and behaviours.

Finally, thank you to all the readers who have taken the time to delve into this book. Your curiosity and engagement with this topic has made the writing process an incredibly rewarding one.

Index

Powered by Penguin

Looking for more great reads, exclusive content and book giveaways?

Subscribe to our weekly newsletter.

Scan the QR code or visit penguin.com.au/signup